ATTACKING PROBABILITY AND STATISTICS PROBLEMS

David S. Kahn

Dover Publications, Inc., Mineola, New York

Bibliographical Note

Attacking Probability and Statistics Problems is a new work, first published by Dover Publications, Inc., in 2016.

Library of Congress Cataloging-in-Publication Data

Names: Kahn, David S.
Title: Attacking probability and statistics problems / David S. Kahn.
Description: Mineola, New York : Dover Publications, Inc., 2016.
Identifiers: LCCN 2015047548| ISBN 9780486801445 | ISBN 0486801446
Subjects: LCSH: Mathematical statistics—Study and teaching. | Probabilities—Study and teaching.
Classification: LCC QA276.18 .K34 2016 | DDC 519.5—dc23 LC record available at http://lccn.loc.gov/2015047548

Manufactured in the United States
80144601 2016
www.doverpublications.com

To the Reader:

Welcome to Probability and Statistics! In most of the mathematics that you have probably studied before this, you usually are given a problem for which you try to find the exact answer. One of the challenging aspects of Probability and Statistics is that there often is not necessarily an exact answer to a given problem. Instead, you may have to find a range of numbers for an answer or you may have to look at some data and decide that your data is not good enough for you to draw a definite conclusion. In Probability and Statistics, we study what might occur or what the data is probably trying to tell us. This can be challenging, but this book will teach you how to attack Probability and Statistics problems and successfully conquer them.

This book is not meant to be an exhaustive study of Probability or Statistics. We will first look at Probability in the first four chapters, focusing on the kind of problems that we will need to be able to solve in order to learn Statistics. While Probability is a fascinating and complex topic, we do not have the space to cover all of the various aspects of the field. Instead, we will look at the basics of Probability, so that we can understand the Statistics portion of the book.

The remaining chapters are devoted to Statistics. They are intended to help you learn to attack the kinds of problems that you will encounter in an elementary Statistics course. Also, much of what is taught in a Statistics course can be calculated using a statistical software program, a spreadsheet program, or even a calculator. Although we will show you how to calculate a variety of statistics, we leave the more cumbersome and obscure calculations for you to carry out with your computer. Instead, it is more important to understand what the statistics mean than to be able to do the calculations themselves (which can be tedious).

We have organized this book so that you can proceed from one topic to another, or you can jump to the topics that you want to work on. To be truthful, you will probably find that you need to go through all of the units if you truly wish to be able to attack Probability and Statistics. The book is divided into 12 units, each of which will teach you what you need to know to do well in that topic. This book focuses on the essentials and how to master the problems. We suggest that you read through each unit completely, do all of the exercises, and complete all of the practice problems. Each example and problem has a complete explanation to help you understand how to solve the problem correctly. There are many good textbooks on Probability and Statistics, and after you have worked through a unit, you may want to refer to a textbook for further practice on that unit's topic.

Probability and Statistics is a fascinating area of mathematics. So much of what you will encounter in your daily lives involves both topics. After you have gone through this book, you will be able to do well in your Statistics course. You will also have a better understanding of the statistics that you will read and hear about. Are you ready? Then it's time to Attack Probability and Statistics!

Acknowledgments

First of all, I would like to thank my editor, Janet Kopito, for her patience and meticulousness. Next, I would like to thank Marisa Bruno for working through all of the problems and double-checking my calculations. I would also like to thank the copy editor, Louise Jarvis, and the proofreader, Lynze Greathouse, for finding the many typos that one makes when writing a mathematics book. I owe a lifetime debt to my father, Peter Kahn, and to my dear friend, Arnold Feingold, who encouraged my interest in mathematics and have always been there to guide me through the rough spots. And finally, I would like to thank the very many students whom I have taught and tutored, who never hesitate to correct me when I am wrong, and who provide the fulfillment that I so deeply derive from teaching math.

Table of Contents

ATTACKING PROBABILITY AND STATISTICS PROBLEMS

UNIT ONE

Basic Probability

In this unit, we are going to learn some of the basics of probability. Statistics, as we will learn, is a way of describing and analyzing events that have already occurred. Probability, on the other hand, is a way of predicting events that might occur. In order to understand the probability that a specific event might occur, we need to know about how often that event might occur relative to how often related events might occur. For example, if we roll a die, how often will the number 1 occur compared to how often any of the numbers from 1 to 6 might occur. Given that there are six numbers on a die and, assuming that it is an honest die, any of those numbers could occur when we roll the die, we expect the number 1 to occur $\frac{1}{6}$ of the time. Suppose, instead, that we had a 10-sided die, with the numbers 1 to 10 on it. If we roll that die, we would expect the number 1 to occur $\frac{1}{10}$ of the time. Going back to our regular die, how often would we expect an even number to occur? Well, there are 3 even numbers, {2, 4, 6}, and 3 odd numbers, {1, 3, 5}. When we roll the die, we would expect an even number to occur $\frac{3}{6}$ or $\frac{1}{2}$ of the time.

This leads us to our first rule:

Rule #1: The probability that an event will occur is

$$\frac{\text{number of ways the specific event can occur}}{\text{total number of possible outcomes}}.$$

Note that when we use the term *event*, what we mean is a roll of the die, a toss of the coin, etc. The denominator, or total number of possible outcomes is referred to as the *sample space*. For example, if we roll a 6-sided die, the sample space is the set of numbers that we can roll, namely {1, 2, 3, 4, 5, 6}. If we toss a coin twice, the sample space is {*HH*, *HT*, *TH*, *TT*}, where *H* represents the head side of a coin (heads) and *T* represents the tail side of a coin (tails). A sample space is very useful for enumerating the possible outcomes when that number is relatively small. We would not want to write out the sample space for tossing 10 coins. There would be $2^{10} = 1024$ elements!

Let's look at some more examples.

Example 1: We roll a 6-sided die. What is the probability that a multiple of 3 will occur?

There are two multiples of 3, namely 3 and 6. There are 6 possible outcomes. Thus, the probability that a multiple of 3 will occur is $\frac{2}{6}$, or $\frac{1}{3}$.

Example 2: We toss a fair coin. What is the probability that we get heads?

There are 2 possible outcomes – heads and tails. There is one way to get heads. Thus, the probability that we will get heads is $\frac{1}{2}$.

By the way, a fair coin means that the chances of getting heads are the same as the chances of getting tails.

These examples are very simple so let's look at some that are a little more complicated. Suppose that we toss a fair coin twice. What is the probability that we get 2 heads?

Let's think about what could happen when we toss a coin twice. We could get heads followed by heads. We could get heads followed by tails. We could get tails followed by heads, or we could get tails followed by tails. To represent these in a convenient shorthand, the possible outcomes are: $\{HH, HT, TH, TT\}$, as noted earlier. Note that there are 4 outcomes and that only one of them is the one that we want, so the probability is $\frac{1}{4}$.

Notice that heads followed by tails is not the same as tails followed by heads. We will see later that sometimes the order in which events happen is important and sometimes it is not.

Example 3: If we toss a fair coin twice, what is the probability that we get at least 1 tail?

If we refer to the previous set of outcomes and sample space, we can see that tails occurs in 3 of the 4 possible outcomes, so the probability is $\frac{3}{4}$.

Example 4: We roll a red die and a green die simultaneously. What is the probability that we roll a 7?

There are many possible outcomes, so let's make a table of them. The numbers are in the order (red, green).

(1, 1)	(2, 1)	(3, 1)	(4, 1)	(5, 1)	(6, 1)
(1, 2)	(2, 2)	(3, 2)	(4, 2)	(5, 2)	(6, 2)
(1, 3)	(2, 3)	(3, 3)	(4, 3)	(5, 3)	(6, 3)
(1, 4)	(2, 4)	(3, 4)	(4, 4)	(5, 4)	(6, 4)
(1, 5)	(2, 5)	(3, 5)	(4, 5)	(5, 5)	(6, 5)
(1, 6)	(2, 6)	(3, 6)	(4, 6)	(5, 6)	(6, 6)

Note that there are 6 ways that we could roll a 7: {(1, 6), (2, 5), (3, 4), (4, 3), (5, 2), (6, 1)}. There are 36 possible outcomes, so the probability of getting a 7 is $\frac{6}{36}$, or $\frac{1}{6}$. You should also notice that, for example, the roll (1, 6) is different from the roll (6, 1). The former consists of the red die showing a 1 and the green die showing a 6, whereas on the latter, the red die shows a 6 and the green die shows a 1. This distinction is important, as we will see later.

Let's look at the table again, this time totaling the result of each roll. We get:

2	3	4	5	6	7
3	4	5	6	7	8
4	5	6	7	8	9
5	6	7	8	9	10
6	7	8	9	10	11
7	8	9	10	11	12

Let's look at the probability of each possible sum. If we roll the two dice, we could get any integer sum ranging from 2 to 12.

What is the probability of rolling a 2? There are 36 possible outcomes and only one of them is a 2, namely a 1 on each die, so $P(2)=\frac{1}{36}$. (By the way, we will use the notation "$P(x)=$" to signify the probability of getting outcome x.)

What is the probability of rolling a 3? There are two ways to get a 3, namely rolling a (1, 2) or a (2, 1), so $P(3)=\frac{2}{36}$.

Let's list them all:

$$P(2)=\frac{1}{36}$$

$$P(3)=\frac{2}{36}$$

$$P(4)=\frac{3}{36}$$

$$P(5)=\frac{4}{36}$$

$$P(6)=\frac{5}{36}$$

$$P(7)=\frac{6}{36}$$

$$P(8)=\frac{5}{36}$$

$$P(9)=\frac{4}{36}$$

$$P(10)=\frac{3}{36}$$

$P(11) = \frac{2}{36}$

$P(12) = \frac{1}{36}$

Let's look at some of the results. The highest probability is that one will roll a 7; the lowest probability is that one will roll either a 2 or a 12. The probabilities start at $\frac{1}{36}$, go up to $\frac{6}{36}$, and then go back down to $\frac{1}{36}$. The probability of rolling a 2 is the same as the probability of rolling a 12; the probability of rolling a 3 is the same as the probability of rolling an 11; the probability of rolling a 4 is the same as the probability of rolling a 10; and so on. There are all sorts of patterns with probabilities and we will see more of these as we explore the subject.

Most important, let's add up the probabilities. We get $\frac{1}{36}+\frac{2}{36}+\frac{3}{36}+\frac{4}{36}+\frac{5}{36}+\frac{6}{36}+\frac{5}{36}+\frac{4}{36}+\frac{3}{36}+\frac{2}{36}+\frac{1}{36}=\frac{36}{36}=1$. This leads us to a very important rule:

The sum of the probabilities of a set of all possible outcomes is 1.

In other words, if we choose an event, say tossing 3 coins, and we list all of the possible outcomes, the sum of the probabilities of these outcomes is always 1. Also:

The probability of any outcome is $0 \leq P(x) \leq 1$.

That is, we cannot get a probability that is negative or greater than 1.

These rules will prove very useful later on. Let's do an example.

Example 5: If we toss 3 coins, what are all of the possible probabilities for getting heads?

We could get the following outcomes: {*HHH, HHT, HTH, HTT, THH, THT, TTH, TTT*}. The probabilities are:

$P(\text{one head}) = \frac{3}{8}$

$P(\text{two heads}) = \frac{3}{8}$

$P(\text{three heads}) = \frac{1}{8}$.

Note that if we add these up, we get $\frac{3}{8}+\frac{3}{8}+\frac{1}{8}=\frac{7}{8}$, but we are supposed to get a total of 1. What are we forgetting? We are forgetting the probability for the outcomes where we get tails only. That is, $P(0 \text{ heads}) = \frac{1}{8}$. Now if we add up the probabilities, we get $\frac{3}{8}+\frac{3}{8}+\frac{1}{8}+\frac{1}{8}=1$.

How did we calculate these probabilities? We listed all of the outcomes and then we looked at the number of ways we could get the outcome we want. This is called the *Counting Principle.* It is an excellent way to calculate probabilities when the number of possible outcomes is small.

Example 6: What is the probability of drawing one queen from an ordinary deck of cards?

There are 52 cards in a deck. Four of the cards are queens, so $P(\text{queen}) = \frac{4}{52} = \frac{1}{13}$.

The Counting Principle is very useful for simple probabilities. However, suppose we are rolling 3 dice. The number of possible outcomes is 216, and it would become unwieldy to write them out all of the time. Thus, there must be other ways to calculate probabilities that are not as tedious. One such way is the *Multiplication Rule.*

The Multiplication Rule says that if $P(A) = x$ and $P(B) = y$, then $P(A \text{ and } B) = xy$.

That is, if we want to find the probability of 2 events occurring in a row, we multiply the 2 probabilities. Furthermore, if we want to find the probability of 3 events occurring in a row, we multiply the 3 probabilities, and so on. Let's do an example.

Example 7: What is the probability of tossing a fair coin twice and getting heads both times?

The probability of getting heads on the first toss is $P(\text{heads on the first toss}) = \frac{1}{2}$. The probability of getting heads on the second toss is also $P(\text{heads on the second toss}) = \frac{1}{2}$. Thus, the probability of getting heads twice is $P(HH) = \frac{1}{2} \cdot \frac{1}{2} = \frac{1}{4}$. By the way, go back to where we listed the 4 outcomes and notice that we could have also used the Counting Principle to get the answer.

Example 8: What is the probability of tossing a coin 6 times in a row and getting heads each time?

Although we could write out all of the possible outcomes, there are 64 of them (we will learn where this comes from later) and it would get tedious. Using the Multiplication Principle, we get $P(HHHHHH) = \frac{1}{2} \cdot \frac{1}{2} \cdot \frac{1}{2} \cdot \frac{1}{2} \cdot \frac{1}{2} \cdot \frac{1}{2} = \left(\frac{1}{2}\right)^6 = \frac{1}{64}$.

Example 9: What is the probability of drawing 2 queens in a row from a deck of playing cards?

This question actually has two possible answers. Why? It depends on what happens after one draws the first card in the deck. Is it replaced in the deck or not? Let's look at the difference.

We draw the first card. The probability of getting a queen is $\frac{4}{52}$. Now, if the card is replaced in the deck, then the probability of getting a queen is again $\frac{4}{52}$, so the probability of getting 2 queens in a row is $\frac{4}{52}\cdot\frac{4}{52}=\frac{16}{2704}=\frac{1}{169}$. Suppose instead that after drawing the first card it is not replaced in the deck. Now there are only 51 cards left and only 3 queens (because the first card was a queen), so the probability of getting 2 queens in a row is $\frac{4}{52}\cdot\frac{3}{51}=\frac{12}{2652}=\frac{1}{221}$. Note that this is less probable. This should make intuitive sense because the overall number of cards has been reduced by 1, which is around 2% of the total number of cards. The number of queens has also been reduced by 1, but 1 queen is 25% of the total number of queens.

The first variation of this question is referred to as "with replacement" and the second as "without replacement." As we can see, whether the card is replaced or not makes a difference. Let's do another example.

Example 10: What is the probability of drawing 2 spades in a row (a) with replacement and (b) without replacement?

(a) There are 52 cards in a deck and 13 of them are spades, so the probability of getting a spade on the first draw is $\frac{13}{52}$. After we replace the card, the probability of getting a spade on the second draw is again $\frac{13}{52}$. Therefore, the probability of getting 2 spades in a row is $\frac{13}{52}\cdot\frac{13}{52}=\frac{169}{2704}=\frac{1}{16}$.

(b) This time, the probability of getting a spade on the first draw is again $\frac{13}{52}$. But, after drawing the first card, we do not replace that card in the deck. Now, there are 51 cards left in the deck and 12 spades, so the probability of getting a spade on the second draw is $\frac{12}{52}$. Thus, the probability of getting 2 spades in a row is $\frac{13}{52}\cdot\frac{12}{51}=\frac{156}{2652}=\frac{1}{17}$.

Remember that the sum of the probabilities of a particular set of events occurring is always 1. Thus, if the probability of an event occurring is given as p, then the probability that the event does not occur is $1-p$. This can be a very useful shortcut to finding a probability. Take a look at the next example.

Example 11: If we toss a fair coin 4 times, what is the probability of getting at least 1 head?

We could find the probabilities of getting 1, 2, 3, or 4 heads, and then add up the probabilities. Or, we could find the probability that we get 0 heads, and subtract that from 1. Why can we do that? Because there are only 5 possibilities, 0, 1, 2, 3, or 4 heads. The sum of the probabilities of these 5 outcomes must be 1. So, if we subtract the probability of 0 heads from 1, we will be left with the sum of the other 4 probabilities. What is the probability of getting 0 heads? The probability that we do not get heads (that is, tails) on any toss is $\frac{1}{2}$, so the probability of getting 0 heads on 4 tosses is $\frac{1}{2}\cdot\frac{1}{2}\cdot\frac{1}{2}\cdot\frac{1}{2}=\frac{1}{16}$. Thus, the probability of getting at least 1 heads is $1-\frac{1}{16}=\frac{15}{16}$. Work this out for yourself to see that it is true.

Let's do another example.

Example 12: We toss a fair coin 10 times. What is the probability that we get at least 1 head?

We could find the probabilities of getting 1, 2, 3, 4, 5, 6, 7, 8, 9, and 10 heads and then add them up. However, it would be much simpler to find the probability of getting zero heads and subtracting that from 1. The probability of getting 0 heads is $\underbrace{\frac{1}{2}\cdot\frac{1}{2}\cdot\frac{1}{2}\cdot\ldots\frac{1}{2}}_{10\text{ times}}=\left(\frac{1}{2}\right)^{10}=\frac{1}{1024}$. Thus, the probability of getting at least one head is $1-\frac{1}{1024}=\frac{1023}{1024}$.

Let's do some practice problems.

Practice Problems

Practice Problem 1: If we roll a 6-sided die, what is the probability that we will get a prime number?

Practice Problem 2: If we toss a fair coin 3 times, what is the probability of getting 3 tails?

Practice Problem 3: If we toss a red and a green die simultaneously, what is the probability of rolling either a 7 or an 11?

Practice Problem 4: If we toss a red and a green die simultaneously, what is the probability of *not* rolling either a 7 or an 11?

Practice Problem 5: A jar contains 6 red, 8 blue, and 4 green marbles. What is the probability of drawing two red marbles in a row (a) with replacement; (b) without replacement?

Practice Problem 6: We draw 2 cards in a row, without replacement, from a standard deck of cards. What is the probability that they are both aces?

Practice Problem 7: We draw 3 cards in a row, without replacement, from a standard deck of cards. What is the probability that we get 3 hearts?

Practice Problem 8: We toss a coin 5 times. What is the probability that we get at least 1 tail?

Practice Problem 9: If the probability of having a male baby is 0.51, what is the probability of a family having three babies, all of them male?

Practice Problem 10: If the probability of having a male baby is 0.51, what is the probability of a family having three babies, at least one of which is male?

Solutions to Practice Problems

Solution to Practice Problem 1: *If we roll a 6-sided die, what is the probability that we will get a prime number?*

The possible rolls are 1, 2, 3, 4, 5, and 6. Of these rolls, 2, 3, and 5 are prime numbers. So the probability of rolling a prime number is $\frac{3}{6} = \frac{1}{2}$.

Solution to Practice Problem 2: *If we toss a fair coin 3 times, what is the probability of getting 3 tails?*

The probability of getting tails when we toss a fair coin is $\frac{1}{2}$, so the probability of getting 3 tails is $\frac{1}{2} \cdot \frac{1}{2} \cdot \frac{1}{2} = \frac{1}{8}$.

Solution to Practice Problem 3: *If we toss a red and a green die simultaneously, what is the probability of rolling either a 7 or an 11?*

When we roll the two dice, there are many possible outcomes, so let's make a table of them. The numbers are in the order (red, green). We could get:

(1, 1)	(2, 1)	(3, 1)	(4, 1)	(5, 1)	(6, 1)
(1, 2)	(2, 2)	(3, 2)	(4, 2)	(5, 2)	(6, 2)
(1, 3)	(2, 3)	(3, 3)	(4, 3)	(5, 3)	(6, 3)
(1, 4)	(2, 4)	(3, 4)	(4, 4)	(5, 4)	(6, 4)
(1, 5)	(2, 5)	(3, 5)	(4, 5)	(5, 5)	(6, 5)
(1, 6)	(2, 6)	(3, 6)	(4, 6)	(5, 6)	(6, 6)

If we total the results, we get:

2	3	4	5	6	7
3	4	5	6	7	8
4	5	6	7	8	9
5	6	7	8	9	10
6	7	8	9	10	11
7	8	9	10	11	12

There are 6 ways to roll a 7 and 2 ways to roll an 11, giving a total of 8 possibilities. There are 36 possible rolls, so the probability of rolling either a 7 or an 11 is $\frac{8}{36} = \frac{2}{9}$.

Solution to Practice Problem 4: *If we toss a red and a green die simultaneously, what is the probability of not rolling either a 7 or an 11?*

As we saw in the previous problem, the probability of rolling either a 7 or an 11 is $\frac{2}{9}$. We know that the sum of the probabilities of all possible rolls is 1, so the probability of *not* rolling a 7 or an 11 is $1-\frac{2}{9}=\frac{7}{9}$.

Solution to Practice Problem 5: *A jar contains 6 red, 8 blue, and 4 green marbles. What is the probability of drawing 2 red marbles in a row (a) with replacement; (b) without replacement?*

(a) There are 18 marbles in the jar, so the probability of drawing a red marble on the first draw is $\frac{6}{18}$. Now, we replace the marble after we have drawn it, so the probability of drawing a red marble on the second draw is again $\frac{6}{18}$. Thus, the probability of drawing 2 red marbles in a row is $\frac{6}{18}\cdot\frac{6}{18}=\frac{1}{9}$.

(b) The probability of drawing a red marble on the first draw is $\frac{6}{18}$. Because we do not replace the marble after drawing it, there are now only 5 red marbles in the jar and a total of only 17 marbles. The probability of drawing a red marble on the second draw is $\frac{5}{17}$. Thus, the probability of drawing 2 red marbles in a row is $\frac{6}{18}\cdot\frac{5}{17}=\frac{5}{51}$.

Solution to Practice Problem 6: *We draw 2 cards in a row, without replacement, from a standard deck of cards. What is the probability that they are both aces?*

There are 52 cards in a standard deck and there are 4 aces, so the probability that the first card is an ace is $\frac{4}{52}$. After we draw an ace, there are only 51 cards left in the deck and only 3 aces, so the probability that the second card is an ace is $\frac{3}{51}$. Thus, the probability of drawing both aces is $\frac{4}{52}\cdot\frac{3}{51}=\frac{1}{221}$.

Solution to Practice Problem 7: *We draw 3 cards in a row, without replacement, from a standard deck of cards. What is the probability that we get 3 hearts?*

There are 52 cards in a standard deck and 13 of them are hearts, so the probability that the first card is a heart is $\frac{13}{52}$. After we draw the first heart, there are only 51 cards left in the deck and only 12 hearts, so the probability that the second card is a heart is $\frac{12}{51}$. Now, after drawing 2 hearts, there are only 50 cards left in the deck and only 11 hearts, so the probability that the third card is a heart is $\frac{11}{50}$. Thus, the probability that all 3 cards are hearts is $\frac{13}{52}\cdot\frac{12}{51}\cdot\frac{11}{50}=\frac{1716}{132600}=\frac{11}{850}$.

Solution to Practice Problem 8: *We toss a coin 5 times. What is the probability that we get at least 1 tail?*

We need to find the probabilities that we get 1, 2, 3, 4, or 5 tails and add them up. It would be much simpler to find the probability that we get no tails and subtract that probability from 1. The probability that we do not get tails (that is, we get heads) on any toss, is $\frac{1}{2}$, so the probability that we get 5 heads in a row is $\frac{1}{2}\cdot\frac{1}{2}\cdot\frac{1}{2}\cdot\frac{1}{2}\cdot\frac{1}{2}=\left(\frac{1}{2}\right)^5=\frac{1}{32}$. Thus, the probability that we get at least one tail is $1-\frac{1}{32}=\frac{31}{32}$.

Solution to Practice Problem 9: *If the probability of having a male baby is 0.51, what is the probability of a family having three babies, all of them male?*

Because the probability of having a male baby is 0.51, the probability of having three male babies in a row is $(0.51)\,(0.51)\,(0.51) = (0.51)^3 = 0.132651$.

Solution to Practice Problem 10: *If the probability of having a male baby is 0.51, what is the probability of a family having three babies, at least one of which is male?*

We need to find the probabilities of having 1, 2, or 3 male babies and add them up. It would be simpler to find the probability of having no male babies, and subtracting that probability from 1. If the probability of having a male baby is 0.51, then the probability of not having a male baby is $1 - 0.51 = 0.49$. Then the probability of having three babies, none of them male, is $(0.49)\,(0.49)\,(0.49) = (0.49)^3 = 0.117649$. Therefore, the probability of having at least one male baby out of 3 babies is $1 - 0.117649 = 0.882351$.

UNIT TWO

Permutations and Combinations

Now that we have seen the basics of probability, we will learn how to do some more difficult problems. One tool that we will need to attack these problems involves permutations and combinations. Remember that determining the probability that an event occurs requires us to count the number of ways that the event can occur and divide it by the number of total possible occurrences of that event. When we did this in Unit One, we did not pay much attention to the order of the events. In this unit, we will learn when the order of events is important and see its effect on the probability of that event.

We need to figure out ways to order events. Each set of ordered events is called an *arrangement* (we will learn other terms shortly), and we can systematically figure out how to count these arrangements. For example, how many ways can we arrange the letters A and B? There are only two ways–AB and BA. This is not very challenging, but, if we increase the number of letters, the number of arrangements will increase rapidly.

Now, let's see how many ways we can arrange the letters A, B, and C. They are ABC, ACB, BAC, BCA, CAB, and CBA. That is, there are $3 \cdot 2 \cdot 1 = 6$ arrangements. (Note how we counted the arrangements. We listed A first and then arranged B and C. Next, we listed B first and then arranged A and C. Finally, we listed C first and then arranged A and B. We recommend this approach for when we want to count larger arrangements.)

Let's do an example of a larger set of arrangements.

Example 1: Suppose that we have 4 people and 4 seats for them to sit in. The seats are in a row and are designated by the letters A, B, C, and D. How many ways can the four people be arranged in the seats?

The ways that the people can sit (in other words, the arrangements of the letters

$$A,\ B,\ C, \text{ and } D \text{ is } \left\{ \begin{array}{cccccc} ABCD, & ABDC, & ACBD, & ACDB, & ADBC, & ADCB \\ BACD, & BADC, & BCAD, & BCDA, & BDAC, & BDCA \\ CABD, & CADB, & CBAD, & CBDA, & CDAB, & CDBA \\ DABC, & DACB, & DBAC, & DBCA, & DCAB, & DCBA \end{array} \right\},$$

giving a total of 24 arrangements.

That was a lot of work. It's a good thing that we didn't have to seat 10 people! There has to be a better way of enumerating the arrangements. Let's think. Suppose that the first person can sit in any of the 4 seats. Once that person sits, the next person has only 3 possible seats to choose among. Now that person sits. The next person has to choose between only 2 seats. The last person has no choice. The number of arrangements is thus $4 \cdot 3 \cdot 2 \cdot 1 = 24$. Let's do this again with another example.

Example 2: Suppose that we have 6 people and 6 seats for them to sit in. The seats are in a row and are designated by the letters *A, B, C, D, E,* and *F.* How many ways can the 6 people be arranged in the seats?

Using the system that we just developed above, the first person has 6 seats to choose among; the second person has 5 seats to choose among; the third person has 4 seats to choose among; etc. The number of arrangements is, thus, $6 \cdot 5 \cdot 4 \cdot 3 \cdot 2 \cdot 1 = 720$.

As you can see, the number of arrangements grows very quickly each time we add another person. For example, if we arrange 10 people in 10 seats, there are 3,638,800 ways to seat them!

It is cumbersome to write out the numbers in these arrangements, so mathematicians have invented a function that expresses this total of arrangements.

The number of ways that n objects can be ordered in a set of n is called n *factorial*, and is represented by $n!$. That is $n! = n\ (n-1)\ (n-2)\ (n-3) \ldots 1$. For example, as we saw above, the number of ways that we can arrange the four letters *A, B, C,* and *D* is $4! = 4 \cdot 3 \cdot 2 \cdot 1 = 24$.

Most calculators are able to calculate factorials but it is useful to do a few of them by hand to get a feel for the numbers and the speed with which they grow.

$1! = 1$
$2! = 2 \cdot 1 = 2$
$3! = 3 \cdot 2 \cdot 1 = 6$
$4! = 4 \cdot 3 \cdot 2 \cdot 1 = 24$
$5! = 5 \cdot 4 \cdot 3 \cdot 2 \cdot 1 = 120$
$6! = 6 \cdot 5 \cdot 4 \cdot 3 \cdot 2 \cdot 1 = 720$
$7! = 7 \cdot 6 \cdot 5 \cdot 4 \cdot 3 \cdot 2 \cdot 1 = 5040$
$8! = 8 \cdot 7 \cdot 6 \cdot 5 \cdot 4 \cdot 3 \cdot 2 \cdot 1 = 40{,}320$

As we can see, the numbers grow very rapidly and, as you can imagine, it can get very cumbersome to deal with large factorials. Note also a few things:

- First, $1!$ has a value of 1, and $2!$ has a value of 2.
- Second, we only calculate the factorial of non-negative integers.
- Finally, we define $0! = 1$. We will see shortly why this is necessary.

Now suppose that we have a group of 5 people, but we only have 3 seats to put them in. How many ways can we seat them? Let's designate the people by the letters *A, B, C, D,* and *E.* We could seat *A, B,* and *C* in the following ways: {*ABC, ACB, BAC, BCA, CBA, CAB*}. Next, we could seat *A, B,* and *D,* then *A, B,* and *E,* and so on. An easier way to calculate this is that there are 5 people who could sit in the first seat, 4 people who could sit in the second seat, and 3 who could sit in the third seat. This gives us $5 \cdot 4 \cdot 3 = 60$ arrangements. Note that this looks like the factorial that we just learned, except that it is missing $2 \cdot 1$. We could have calculated this by taking $5!$ and dividing it by $2!$. We would get $\frac{5!}{2!} = \frac{5 \cdot 4 \cdot 3 \cdot 2 \cdot 1}{2 \cdot 1} = 5 \cdot 4 \cdot 3 = 60$.

Note that we got 2 by taking the total number of people and subtracting the number of seats that we could put them in.

The above kind of arrangement is called a *permutation.* A permutation is the number of ways that r items selected from a total of n possible items and is found by ${}_{n}\mathrm{P}_{r} = \dfrac{n!}{(n-r)!}$.

A permutation requires three very important conditions:

- There must be n objects available, all of which are *different.* The formula does not apply if some of the objects are identical to others.
- We pick r of the n objects *without* replacement.
- The order of the arrangements matters. That is, if the same objects are arranged in a different order, then it makes a different arrangement (for example, *ABC* is different from *ACB*).

(Many calculators can evaluate permutations.)

Note that the first two examples are permutations. Example 1 is the permutation of four people taken four at a time. We get ${}_{4}\mathrm{P}_{4} = \dfrac{4!}{(4-4)!} = \dfrac{4!}{0!} = 4! = 24$. Notice that we had 0! in the denominator. This is why we define 0! = 1; otherwise, we would get nonsense. Similarly, Example 2 is the permutation of six people taken six at a time. We get ${}_{6}\mathrm{P}_{6} = \dfrac{6!}{(6-6)!} = \dfrac{6!}{0!} = 6! = 720$. Let's do another example.

Example 3: We have 4 spots on a committee–president, vice president, treasurer, and secretary. We have 7 people to choose among when selecting a committee. How many different committees could we create?

One way to do this is to say that there are 7 people who could be president. Once we choose a president, then there are 6 people who could be vice-president (if a person could have more than one spot, then it would not be a permutation), and so on. This give us $7 \cdot 6 \cdot 5 \cdot 4 = 840$ permutations of the committee. We now have a fast way to calculate this: ${}_{7}\mathrm{P}_{4} = \dfrac{7!}{(7-3)!} = \dfrac{7!}{4!} = 840$.

Of course, you could have worked this out without using ${}_{n}\mathrm{P}_{r}$ but suppose we had the following example.

Example 4: There are 60 football games on Saturday and there are 12 time slots to show them in. How many different arrangements of football games could be shown?

We could calculate $60 \cdot 59 \cdot 58 \cdot 57 \cdot 56 \cdot 55 \cdot 54 \cdot 53 \cdot 52 \cdot 51 \cdot 50 \cdot 49$, which is time consuming. Or we could use the calculator to find ${}_{60}\mathrm{P}_{12} = 670{,}295{,}125{,}717{,}176{,}960{,}000$. Now that's a big number!

What do we do if some of the objects are identical? The permutations rule changes to deal with the duplicate arrangements. Let's do an example.

Example 5: How many ways can we arrange the letters in the word *MISSISSIPPI*?

There are 11 letters in *MISSISSIPPI* so we might think that there are 11! ways to arrange the letters but the problem is that some of the letters are interchangeable, so we might end up over counting them. We divide the total number of arrangements of the letters by the factorial of each group of duplicate letters (because that is how many ways each of those groups can be arranged). In other words, there are four *S*'s, so there are 4! ways that the *S*'s can be arranged, there are four *I*'s, so there are 4! ways that the *I*'s can be arranged, and there are two *P*'s, so there are 2! ways that the *P*'s can be arranged. Thus the total number of ways that we can arrange the letters in *MISSISSIPPI* is $\frac{11!}{4!4!2!} = 34{,}650$.

The rule is: If there are n objects, with r_1 alike, r_2 alike, ..., r_k alike, then the number of permutations of all n objects is $\frac{n!}{r_1!r_2!...r_k!}$.

Example 6: Suppose we have 15 light bulbs in a row. Four of them are red, 6 are green, 3 are yellow, and 2 are white. How many ways can we arrange the light bulbs?

Using the formula, we get $\frac{15!}{4!6!3!2!} = 6{,}306{,}300$.

Suppose we have three people with the uninteresting names of *A*, *B*, and *C*. As we have seen, there are 6 ways that we can arrange them. But, this presumes that the order of each arrangement matters. That is, the arrangement *ABC* is different from group *ACB*. This is true if, for example, each position in the arrangement is different. For example, the first position could be president of a committee, the second could be the vice president, and the third position could be the secretary. Suppose, however, that the order of each arrangement did not matter, such as when people are going to the movies together. In other words, if you take your friends *ABC* to the movies, that is the same as taking *ACB*. If so, then there would only be one way to arrange *A*, *B*, and *C*. This is an example of a *combination*. In order to find out how many combinations can be made, do the same calculation as for a permutation, and then divide by the factorial of the group size to remove the duplicates. Let's do an example.

Example 7: Suppose we have 3 tickets to a movie and 8 friends to choose among to invite. How many different arrangements of friends can we make?

First, let's find the number of permutations: ${}_8P_3 = \frac{8!}{(8-3)!} = \frac{8!}{5!} = 8 \cdot 7 \cdot 6 = 336$.

Now we divide this number by 3!. Why? Because each permutation can be arranged in 3! ways and we only wish to count each permutation once. That is, we do not want to treat the permutation *ABC* as different from *ACB*, and so on. Now we find $\frac{336}{3!} = 56$.

A *combination* is the number of ways that r items selected from a total of n possible items and is found by ${}_nC_r = \frac{n!}{r!(n-r)!}$.

A combination requires three very important conditions:

- There must be n objects available, all of which are *different*. The formula does not apply if some of the objects are identical to others.
- We pick r of the n objects *without* replacement.
- The order of the arrangements does not matter. That is, if the same objects are arranged in a different order, then it makes a different arrangement (for example, ABC is different from ACB).

(Many calculators can evaluate combinations.)

Notice that the number of combinations of n objects taken r at a time is very similar to the number of permutations of n objects taken r at a time, where a combination is the permutation divided by $r!$. Because $r! \geq 1$, we know that ${}_nC_r \leq {}_nP_r$.

(Note: Many math books use the notation $\binom{n}{r}$ *for a combination of n objects taken r at a time. We will use the* ${}_nC_r$ *notation because it is the notation that most calculators use.)*

Example 8: Suppose that we have 4 different sweaters and 9 friends to offer them to (each gets at most one sweater). How many ways can we offer the sweaters to our friends? Suppose the sweaters were identical. Now how many ways can we offer them?

If we have 9 friends and 4 different sweaters, then the order in which we give them away matters. That is, A gets a blue sweater and B gets a red sweater is different from A gets a red sweater and B gets a blue sweater. This is an example of a permutation, so there are ${}_9P_4 = \frac{9!}{(9-4)!} = \frac{9!}{5!} = 9 \cdot 8 \cdot 7 \cdot 6 = 3024$ ways.

If the sweaters are identical, then the order that we give them away does not matter. That is, A gets a sweater and B gets a sweater is the same as B gets a sweater and then A gets a sweater. This is an example of a combination, so there are ${}_9C_4 = \frac{9!}{4!(9-4)!} = \frac{9!}{4!5!} = \frac{9 \cdot 8 \cdot 7 \cdot 6}{4 \cdot 3 \cdot 2 \cdot 1} = 126$ ways. Got the idea? Let's do another example.

Example 9: Suppose that a sushi platter contains 6 different pieces of sushi and we have 10 different types of sushi to choose among. How many different platters can we make?

Don't be fooled by the fact that the types of sushi are different. It doesn't matter what order we put them on the platter (tuna roll and California roll is the same as California roll and tuna roll). Thus, we need to find the combination of ten types of sushi taken six at a time: ${}_{10}C_6 = \frac{10!}{6!(10-6)!} = \frac{10!}{6!4!} = \frac{10 \cdot 9 \cdot 8 \cdot 7 \cdot 6 \cdot 5}{6 \cdot 5 \cdot 4 \cdot 3 \cdot 2 \cdot 1} = 210$.

What do we do about an arrangement that is neither a combination or a permutation? What we do is we figure out the possible number of elements in each location of the arrangement and multiply them together. Let's do an example.

Example 10: A phone number has 7 digits. The first digit cannot be a 0 or 1. How many different phone numbers can be made?

Note that this is not a permutation because, even though the order of the digits matters (Think about it!), a permutation requires that the elements be selected without replacement. That is, if the phone number was a permutation, we would not be able to repeat digits. So, what we do is figure out how many digits we can choose to go in each position of the phone number and multiply the choices together. There are 10 digits (the numbers 0 through 9), and any of them can go in the first position except a 0 or 1, so there are 8 choices for the first position in the phone number. We can put any of the 10 digits in the remaining 6 positions. This gives us: $\underline{8}\cdot\underline{10}\cdot\underline{10}\cdot\underline{10}\cdot\underline{10}\cdot\underline{10}\cdot\underline{10}=8{,}000{,}000$. Now we know why there are area codes!

Example 11: A phone number consists of a 3-digit area code and a 7-digit number. The area code's first digit cannot be a 0 or 1, and the second digit must be a 0 or 1. In the 7-digit number, the first digit cannot be a 0 or 1 (see Example 10). How many different phone numbers can be made?

Let's look at the area code first. The first digit cannot be a 0 or 1, so there are 8 choices for the first digit. The second digit must be a 0 or 1, so there are 2 choices for the second digit. The third digit can be any of the 10 digits. This gives us: $8 \cdot 2 \cdot 10 = 160$. As we saw in Example 10 above, there are 8,000,000 choices for the remaining seven digits of the phone number. When we multiply the two, we get $160 \cdot 8{,}000{,}000 = 1{,}280{,}000{,}000$. You may think that this is a lot of phone numbers but, believe it or not, the United States used this system until the mid-1980s and ran out of phone numbers. Now, the second digit can be any number, so there are $800 \cdot 8{,}000{,}000 = 6{,}400{,}000{,}000$ available numbers.

Of course, we will need to use permutations and combinations to calculate probabilities. We will see how to do so in the next unit.

Practice Problems

Practice Problem 1: In order to win a certain lottery, one must choose 6 correct numbers from the numbers 1 through 46. The numbers can be chosen in any order. How many different combinations of lottery numbers are there?

Practice Problem 2: A basketball team has 5 positions–point guard, shooting guard, small forward, power forward, and center. There are 12 people to choose among. How many different ways can a team be chosen?

Practice Problem 3: We have 5 tickets to a basketball game and 12 friends to offer them to. The seats are not numbered. How many ways could we offer the tickets (each friend gets at most one ticket)?

Practice Problem 4: How many ways can we arrange the letters in the word *STATISTICS*?

Practice Problem 5: A license plate consists of 3 letters followed by 4 digits. How many different license plates can be made?

Practice Problem 6: A true-false quiz has 10 questions. How many different sets of true-false answers can be created?

Practice Problem 7: A uniform consists of a hat, a jacket, and pants. There are 4 choices of color for the hat, 3 colors for the jacket, and 5 colors for the pants. How many different uniforms arrangements are possible?

Practice Problem 8: You have 6 prizes to give away to 11 possible recipients. How many different ways can the prizes be given away if (a) the prizes are all different; (b) the prizes are all the same?

Practice Problem 9: A class consists of 16 students but, unfortunately, there are only 12 seats in the classroom. (a) How many different ways can the students be seated? (b) What if the seats are numbered?

Practice Problem 10: A password consists of 8 characters, each of which can be a digit, a letter of the alphabet, or one of the special characters *, !, ?, /, @, and &. How many different passwords are possible if (a) no characters can be repeated; (b) any character can be repeated; (c) any character can be repeated but the first and last characters are not special characters?

Solutions to Practice Problems

Solution to Practice Problem 1: *In order to win a certain lottery, one must choose 6 correct numbers from the numbers 1 through 46. The numbers can be chosen in any order. How many different combinations of lottery numbers are there?*

The order of the lottery numbers does not matter, so this is a combination of 6 numbers out of a possible 46. We get ${}_{46}C_6 = \frac{46!}{6!(46-6)!} = \frac{46!}{6!40!} = 9{,}366{,}819$ combinations. We also could have found the number of arrangements of the six numbers by evaluating $\underline{46} \cdot \underline{45} \cdot \underline{44} \cdot \underline{43} \cdot \underline{42} \cdot \underline{41} = 6{,}744{,}109{,}680$ and dividing by the duplicates, 6!, to get $\frac{6{,}744{,}109{,}680}{6!} = 9{,}366{,}819$. Note that this second approach is more cumbersome and would be rather difficult with bigger numbers.

Solution to Practice Problem 2: *A basketball team has 5 positions–point guard, shooting guard, small forward, power forward, and center. There are 12 people to choose among. How many different ways can a team be chosen?*

Here, each player has a different position, so this is a permutation of 5 numbers out of a possible 12. We get ${}_{12}P_5 = \frac{12!}{(12-5)!} = \frac{12!}{7!} = 95{,}040$. We also could have found the number of arrangements of the five players by evaluating $\underline{12} \cdot \underline{11} \cdot \underline{10} \cdot \underline{9} \cdot \underline{8} = 95{,}040$.

Solution to Practice Problem 3: *We have 5 tickets to a basketball game and 12 friends to offer them to. The seats are not numbered. How many ways could we offer the tickets (each friend gets at most one ticket)?*

Here, the order that we give out the tickets does not matter because the seats are interchangeable, so this is a combination of 5 numbers out of a possible 12. We get ${}_{12}C_5 = \frac{12!}{5!(12-5)!} = \frac{12!}{5!7!} = 792$.

If we compare Practice Problems 2 and 3, we see that one is a permutation of 5 numbers out of a possible 12 and that the other is a combination of 5 numbers out of a possible 12. Note how many more permutations there are than combinations. This is because each group of 5 numbers can be arranged in 5! = 120 different ways.

Solution to Practice Problem 4: *How many ways can we arrange the letters in the word STATISTICS?*

This problem is like the *MISSISSIPPI* problem (Example 5). That is, it is a permutation where we divide out the duplicates. There are 10 letter in the word *STATISTICS*, with three *S*'s, three *T*'s, and two *I*'s. We get $\frac{10!}{3!3!2!} = 50{,}400$.

Solution to Practice Problem 5: *A license plate consists of 3 letters followed by 4 digits. How many different license plates can be made?*

There are 26 letters in the alphabet and this problem does not say anything about duplicate letters, so there are $26 \cdot 26 \cdot 26 = 26^3 = 17{,}576$ ways to arrange the letters. There are 10 digits and, again, the problem does not say anything about duplicates, so there are $10 \cdot 10 \cdot 10 \cdot 10 = 10^4 = 10{,}000$ ways to arrange the digits. Therefore, in total, there are $17{,}576 \cdot 10{,}000 = 175{,}760{,}000$ possible license plates.

Solution to Practice Problem 6: *A true-false quiz has 10 questions. How many different sets of true-false answers can be created?*

There are only two possibilities for each answer–true or false. There are 10 questions, so the total number of sets of answers is $2 \cdot 2 \cdot 2 \cdot 2 \cdot 2 \cdot 2 \cdot 2 \cdot 2 \cdot 2 \cdot 2 = 2^{10} = 1024$ possible sets.

Solution to Practice Problem 7: *A uniform consists of a hat, a jacket, and pants. There are 4 choices of color for the hat, 3 colors for the jacket, and 5 colors for the pants. How many different uniforms arrangements are possible?*

There are 4 colors for the hat, 3 colors for the jacket, and 5 colors for the pants, so we get $4 \cdot 3 \cdot 5 = 60$ different arrangements.

Solution to Practice Problem 8: *You have 6 prizes to give away to 11 possible recipients. How many different ways can the prizes be given away if (a) the prizes are all different; (b) the prizes are all the same?*

(a) The prizes are all different, so this is a permutation of 6 numbers out of a possible 11. We get ${}_{11}P_6 = \dfrac{11!}{(11-6)!} = \dfrac{11!}{5!} = 332{,}640$ different ways to give away the prizes.

(b) The prizes are all the same, so this is a combination of 6 numbers out of a possible 11. We get ${}_{11}C_6 = \dfrac{11!}{6!(11-6)!} = \dfrac{11!}{6!5!} = 462$ different ways to give away the prizes.

Solution to Practice Problem 9: *A class consists of 16 students but, unfortunately, there are only 12 seats in the classroom. (a) How many different ways can the students be seated? (b) What if the seats are numbered?*

(a) If the seats are all the same, then this is a combination of 12 numbers out of a possible 16. We get ${}_{16}C_{12} = \dfrac{16!}{12!(16-12)!} = \dfrac{16!}{12!4!} = 1820$ different ways to seat the students.

(b) If the seats are numbered, then the order in which they are seated matters, so this is a permutation of 12 numbers out of a possible 16. We get ${}_{16}P_{12} = \dfrac{16!}{(16-12)!} = \dfrac{16!}{4!} = 871{,}782{,}912{,}000$ different ways to seat the students.

Solution to Practice Problem 10: *A password consists of 8 characters, each of which can be a digit, a letter of the alphabet, or one of the special characters *, !, ?, /, @, and &. How many different passwords are possible if (a) no characters can be repeated; (b) any character can be repeated; (c) any character can be repeated but the first and last characters are not special characters?*

(a) There are 10 digits, 26 letters, and 8 special characters from which to choose, for a total of 44 possible choices. The order of the characters in a password matters (think about it!). Because no character can be repeated, the password is a permutation of 8 characters out of a possible 44. We get ${}_{44}P_8 = \dfrac{44!}{(44-8)!} = \dfrac{44!}{36!} = 7{,}146{,}019{,}520{,}640$ possible passwords.

(b) If any character can be repeated, then this is not a permutation or a combination. There are 44 possible characters for each position in the password, so there are $44 \cdot 44 \cdot 44 \cdot 44 \cdot 44 \cdot 44 \cdot 44 \cdot 44 = 44^8 = 14{,}048{,}223{,}625{,}216$ possible passwords.

(c) If any character can be repeated, then this is not a permutation or a combination. There are 44 possible characters for each position in the password, except that there are only 36 choices for the first and last positions (because those can not be special characters). Thus, there are $36 \cdot 44 \cdot 44 \cdot 44 \cdot 44 \cdot 44 \cdot 44 \cdot 36 = 36^2 \cdot 44^6 = 9{,}404{,}182{,}757{,}376$ possible passwords.

UNIT THREE

More Probability

As we have seen, the probability that an event will occur is found by taking the number of ways that it can occur and dividing that number by the total possible number of outcomes in the sample space. Now that we have learned about permutations and combinations, we can do some more complicated types of probability problems.

Let's look again at a standard deck of cards. Remember (or you might not know) that a deck of cards contains four suits – spades and clubs, which are black, and hearts and diamonds, which are red. Within each suit, there are thirteen cards–2, 3, 4, 5, 6, 7, 8, 9, 10, Jack, Queen, King, and Ace. This gives a total of 52 cards. Thus, the probability of getting a red card is $\frac{26}{52}=\frac{1}{2}$. Of course, things can get more complicated.

Example 1: What is the probability of getting a two of a kind (a pair) in a 5-card hand, with the other three cards having different values and not matching the pair?

We are looking for the probability of a hand of the form *AABCD*, where *A*, *B*, *C*, and *D* are all different kinds of cards. First, let's find the total number of ways to choose 5 cards from the deck of 52 cards. This is simply the combination ${}_{52}C_5=\frac{52!}{5!(52-5)!}=2{,}598{,}960$. This will be the denominator of our probability (the sample space).

The numerator takes a little more work. First, there are 13 ways to choose the value of the pair (2, 3, 4,..., *A*). Next, there are ${}_4C_2 = 6$ ways to choose that pair (for example, Jacks), from each set of four cards in the deck. There are also ${}_{12}C_3 = 220$ ways to choose the 3 other different cards. Finally, there are ${}_4C_1 = 4$ ways of choosing each of those 3 cards, giving us $4 \cdot 4 \cdot 4 = 4^3$. If we multiply these together, we get $13 \cdot {}_4C_2 \cdot {}_{12}C_3 \cdot 4^3 = 1{,}098{,}240$ total hands that contain a pair.

Thus, the probability of getting two of a kind is

$$\frac{13 \cdot {}_4C_2 \cdot {}_{12}C_3 \cdot 4^3}{{}_{52}C_5}=\frac{1{,}098{,}240}{2{,}598{,}960}=0.423.$$

That was much harder than the types of probability that we have seen so far! Let's try another example.

Example 2: What is the probability of getting a three of a kind (a triple) in a 5-card hand, with the other two cards having different values and not matching the pair?

Here, we are looking for the probability of a hand of the form *AAABC*, where *A*, *B*, and *C* are all different kinds of cards. First, let's find the total number of ways to choose 5 cards from the deck of 52 cards. This is simply the combination ${}_{52}C_5 = 2{,}598{,}960$. This will be the denominator of our probability (the sample space).

Now the numerator takes a little more work. First, there are 13 ways to choose the value of the triple (2, 3, 4, *A*). Next, there are ${}_4C_3 = 4$ ways to choose that triple from the four in the deck. Next, there are ${}_{12}C_2 = 66$ ways to choose the two other different cards. Finally, there are ${}_4C_1 = 4$ ways of choosing each of those two cards, giving us $4 \cdot 4 = 4^2$. If we multiply these together, we get $13 \cdot {}_4C_3 \cdot {}_{12}C_2 \cdot 4^2 = 54{,}912$ total hands that contain a pair.

Thus, the probability of getting a triple is

$$\frac{13 \cdot {}_4C_3 \cdot {}_{12}C_2 \cdot 4^2}{{}_{52}C_5} = \frac{54{,}912}{2{,}598{,}960} = 0.0211 .$$

Got the idea? Let's do a different kind of example.

Example 3: A couple plans to have six children. What is the probability that they will have 3 boys and 3 girls?

First, let's figure out how many different gender sequences are possible. For example, the couple could have *BGGBBB* or *BBBBBG* or *GGBBGG*, etc. There are two possibilities for the child's gender and there are six positions in the sequence, so the total number of possible gender sequences is $2 \cdot 2 \cdot 2 \cdot 2 \cdot 2 \cdot 2 = 2^6 = 64$.

Next, let's figure out how many different gender sequences of 3 boys and 3 girls are possible. We could think of this as how many ways can we arrange the letters *GGGBBB*. This is a combination of 3 out of a possible 6, or ${}_6C_3 = 20$.

Thus, the probability of having 3 boys and 3 girls is $\frac{{}_6C_3}{2^6} = \frac{20}{64} = \frac{5}{16}$.

Before we deal with the next kind of probability problems, we will need to learn some more terminology. First, a *compound event* is any event combining two or more individual events. For example, suppose we were selling ice cream cones. Someone could get a vanilla cone, a chocolate cone, or a vanilla *and* chocolate cone, or a cone with *neither* flavor. In these kinds of situations, we have to be careful about double counting events. Let's figure out a rule by first doing an example.

Example 4: An ice cream stand sells 120 ice cream cones. Sixty-five cones are vanilla, 40 cones are chocolate, and 30 cones are vanilla *and* chocolate. How many cones are neither vanilla nor chocolate?

First, add up the three numbers of cones. We get 65 + 40 + 30 = 135 cones. How is that possible? There were only 120 cones. Simple, we double counted some of the cones. If someone had a vanilla and chocolate cone, that person would be counted as having both a vanilla cone and a chocolate cone. We want to count that person only once. This is easy to visualize using a *Venn Diagram*, which consists of circles that represent the number of outcomes of each event.

Figure 1

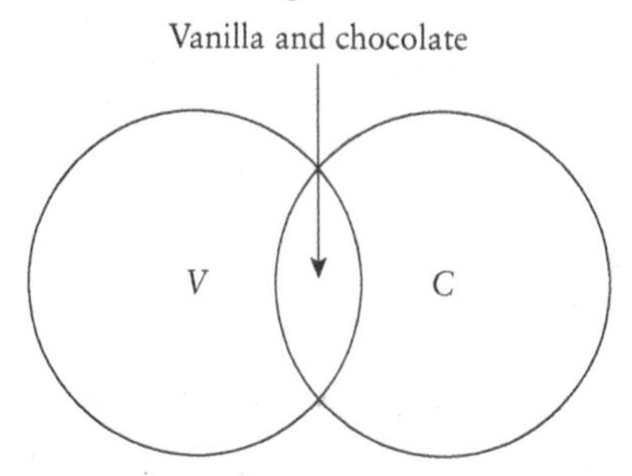

Here, we have a circle labeled V (for vanilla), a circle labeled C, (for chocolate), a region where the circles overlap (vanilla and chocolate), and the outside of the circles (neither vanilla nor chocolate). If we put 30 in the overlapping region (for the 30 vanilla and chocolate cones), then there are $65 - 30 = 35$ cones that must be only vanilla. Similarly, there are $40 - 30 = 10$ cones that must be only chocolate. If we add up the 3 regions, we get $35 + 30 + 10 = 75$ cones. There are a total of 120 cones, so there are $120 - 75 = 45$ cones that are neither vanilla nor chocolate. Let's do another one.

Example 5: There are 250 students in the eleventh grade. One hundred eighty students take Spanish, 90 take Italian, and 40 take both. What is the probability that a student takes neither Spanish nor Italian?

Let's make a Venn Diagram of the situation:

Figure 2

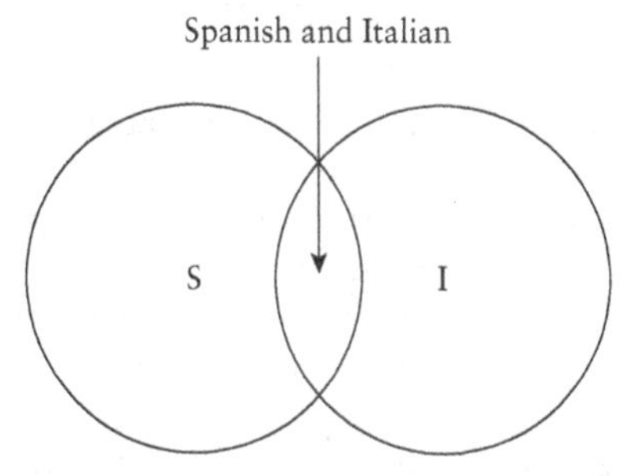

There are 40 students in the overlapping region (Spanish and Italian), so there are $180 - 40 = 140$ students who take only Spanish. There are $90 - 40 = 50$ students who take only Italian. If we add up the three regions, we get $140 + 40 + 50 = 230$ students. Thus, there are $250 - 230 = 20$ students who take neither Spanish nor Italian, so the probability that a student takes neither language is $\frac{20}{250} = 0.08$.

Suppose we have a Venn Diagram where the two circles are labeled A and B, and the overlap is labeled X (which stands for A and B). If we wish to find how many elements are in A or B, we need to add the three regions together. We have the region X, the region $A - X$, and the region $B - X$. If we add these together, we get $X + (A - X) + (B - X) = A + B - X$. Now, because X is the region A and B, we could rewrite this as $A + B - (A \text{ and } B)$. This gives us the formula $A \text{ or } B = A + B - (A \text{ and } B)$.

This leads us to a very important formula, where each region stands for a probability: $P(A \text{ or } B) = P(A) + P(B) - P(A \text{ and } B)$.

Let's go back to Example 5 and use the formula to find how many students are taking Spanish or Italian. We know that 180 students take Spanish, 90 take Italian, and 40 take both. According to the formula, the number who take Spanish or Italian is $180 + 90 - 40 = 230$. Now, if we want to know how many students take neither Spanish nor Italian, we get $250 - 230 = 20$. That was a lot easier than making a Venn Diagram, wasn't it?

Many times, we will want to look at events that are *mutually exclusive*. These are events that cannot occur together. Another way to think of them is that $P(A \text{ and } B) = 0$. This means that $P(A \text{ or } B) = P(A) + P(B)$. For example, a coin toss can come up heads or tails, but not both. Let's do an example.

Example 6: One hundred women in a doctor's office are tested to see if they are pregnant. A positive result means that the woman is pregnant. A negative result means that she is not pregnant. The test is imperfect, so occasionally we will get an incorrect result. We get the following results:

	Positive Test Result	**Negative Test Result**
Pregnant	52	5
Not Pregnant	2	41

What is the probability of selecting a woman at random and choosing one who is not pregnant or one who has a negative test result?

As we can see, 5 of the women are told that they are not pregnant when they are. Two of the women are told that they are pregnant when they are not. The former is called a *false negative* and the latter is called a *false positive*. As you can imagine, one tries to avoid false positives and false negatives, and we will analyze these kinds of errors later in the book.

There are 41 women who are not pregnant and 5 others who get a negative test result. There are also 2 women who are not pregnant even though they have a positive test result. The total is 48 women who are either not pregnant or have a negative test result. One hundred women are tested, so the probability of choosing a woman who is not pregnant or who gets a negative result is $\frac{48}{100} = 0.48$.

Another way that we could have found the number who are either not pregnant or who have a negative test result is to add the number of women who are not pregnant (43) to the number who have a negative test result (46) and then subtract the number who are both not pregnant and have a negative test result, that is, the number of women who are double counted (41). We get 43 + 46 – 41 = 48.

Note that both ways get us to the right answer. Sometimes one way is faster, sometimes the other. It will usually depend on what information we are given and what we have to figure out.

There is another kind of problem that deals with what is called *Binomial probability*. These deal with outcomes that (1) belong to two categories, such as yes/no, on/off, etc.; (2) where the probability of each outcome remains the same for each trial; and (3) the results of the trials are *independent* (that is, the outcome of any trial does not affect the probability of the outcome of any other trial). Let's do an example and see if we can come up with a formula.

Example 7: A coin is tossed 4 times. What is the probability that the coin comes up heads 2 of the 4 times?

First, let's verify that this is binomial. The coin can either come up heads or tails, the probability is the same each time it is tossed $\left(\frac{1}{2}\right)$ and whether we get heads on a particular toss has nothing to do with the previous toss. So, this is binomial event.

Now, let's figure out the probability of getting heads twice. The probability that the coin is heads on the first toss is $\frac{1}{2}$ and on the second toss is also $\frac{1}{2}$. The probability that the coin comes up tails on the third toss is $\frac{1}{2}$, and the probability that the coin comes up tails on the last toss is $\frac{1}{2}$. Thus, the probability of getting 2 heads followed by 2 tails is $\frac{1}{2} \cdot \frac{1}{2} \cdot \frac{1}{2} \cdot \frac{1}{2} = \frac{1}{16}$. But, this is only the probability that the coin comes up heads twice in a row, followed by tails twice in a row. It could also come up *HTHT*, or *HTTH*, etc. In fact, there are ${}_4C_2 = 6$ different ways that the heads and tails could occur. Each of these will have the same probability $\left(\frac{1}{16}\right)$. So, the probability of getting 2 heads and 2 tails is $6 \cdot \frac{1}{16} = \frac{6}{16} = \frac{3}{8}$.

Note that we took the probability of 1 arrangement of heads and tails, and multiplied it by the number of ways that we can arrange heads and tails in 4 tosses. Here is a more general formula:

The probability that a binomial event will occur r times out of a possible n is $P(r) = {}_nC_r\,(p^r)(q^{n-r})$, where:

n is the total number of events;

r is the number of desired events;

p is the probability that the desired event occurs;

q is the probability that the desired event does *not* occur (that is $1 - p$).

Example 8: Suppose that the probability that it will rain on any given day is 0.3. What is the probability that it will rain on 2 days in any given week?

The probability that it will rain is 0.3, so the probability that it will not rain is 1 – 0.3 = 0.7. There are 7 days in a week, so using the formula, we get ${}_7C_2\,(0.3^2)(0.7^{7-2}) = 21 \cdot (0.09)(0.16807) = 0.318$.

Let's do another example.

Example 9: A light board has 10 light bulbs on it in a row. The bulbs are 95% reliable. That is, the probability that a bulb is defective is 0.05. (a) What is the probability that 2 of them are defective? (b) What is the probability that 3 or more of them are defective?

Again this is a binomial probability problem. Either a bulb is defective or it is not. A defective bulb does not depend on whether any other bulb is defective.

(a) The probability that a bulb is defective is 0.05, so the probability that a bulb is not defective is $1 - 0.05 = 0.95$. There are 10 bulbs in a row, so using the formula, we get ${}_{10}C_2\,(0.05^2)(0.95^{10-2}) = 45 \cdot (0.05^2)(0.95^8) = 0.0746$.

(b) If we want to find the probability that 3 or more are defective, we need to find the probability that 3 of them are defective, that 4 are defective, and so on, up to the probability that 10 are defective. Or we could find the probability that 0, 1, or 2 bulbs (which we have already found) are defective and add them up. Then we subtract this probability from 1.

The probability that 0 bulbs are defective is: ${}_{10}C_0(0.05^0)(0.95^{10}) = 1 \cdot 1 \cdot (0.95^{10}) = 0.599$.

The probability that 1 bulb is defective is: ${}_{10}C_1(0.05^1)(0.95^{10-1}) = 10(0.05)(0.95^9) = 0.315$.

Now we add up the probabilities of 0, 1, or 2 defective bulbs to get $0.599 + 0.315 + 0.0746 = 0.9886$. We now subtract this from one to get $1 - 0.9886 = 0.0114$.

As a real-world note, it may sound good that a bulb will only be defective 5% of the time, but notice that this means that if someone has ten of these bulbs, there is approximately a 40% chance that one or two of them are defective. Let's redo the example with 99% reliability to see what happens.

Example 10: A light board has 10 light bulbs on it in a row. The bulbs are 99% reliable. That is, the probability that a bulb is defective is 0.01. What is the probability that 1 or 2 of them are defective?

The probability that a bulb is defective is 0.01, so the probability that a bulb is not defective is $1 - 0.01 = 0.99$. There are 10 bulbs in a row, so using the formula, we get

$${}_{10}C_2\,(0.01^2)(0.99^{10-2}) = 45 \cdot (0.01^2)(0.99^8) = 0.0042$$

The probability that one is defective is: ${}_{10}C_1\,(0.01^1)(0.99^{10-1}) = 10(0.01)(0.99^9) = 0.0914$.

Now we add up the probabilities that 1 or 2 bulbs are defective to get $0.0042 + 0.0914 = 0.0956$. In other words, even though the bulbs are 99% reliable, if someone has ten of them, there is nearly a 10% chance that 1 or 2 will be defective.

Time to do some practice problems!

Practice Problems

Practice Problem 1: What is the probability of getting 2 pairs in a 5-card hand with the other card not matching either pair?

Practice Problem 2: What is the probability of all of the cards in a 5-card hand being of the same suit (a flush)?

Practice Problem 3: A couple plans to have 8 children. What is the probability that they will have 4 boys and 4 girls?

Practice Problem 4: The serial number on a dollar bill consists of 8 digits in a row. What is the probability that the serial number will consist of 5 zeros and 3 ones?

Practice Problem 5: A college has 400 freshmen. Two hundred forty of the students will take Calculus, 180 of them will take Writing, and 65 will take both. How many will take neither Calculus nor Writing?

Practice Problem 6: Suppose that the probability of getting Disease A is 0.12 and the probability of getting Disease B is 0.05. The probability of getting both is 0.02. What is the probability of getting neither Disease A nor Disease B?

Practice Problem 7: Eighty men in a high school are asked their heights and if they play basketball. We get the following results.

	Plays Basketball	**Does Not Play Basketball**
Six Feet Tall and Over	25	8
Under Six Feet Tall	27	20

What is the probability of selecting a man at random and choosing one who (a) does not play basketball; (b) is under six feet tall and plays basketball; (c) is under six feet tall or plays basketball?

Practice Problem 8: A test consists of multiple choice questions, each having five possible answers (one of which is correct). If a person guesses on 6 questions, (a) what is the probability that the person gets three of them correct? (b) What is the probability that the person gets fewer than 2 questions correct?

Practice Problem 9: The probability that an airline flight will arrive on time at a particular airport is 0.767. On a Friday, 10 flights will arrive at the airport. (a) What is the probability that exactly 8 of them will arrive on time? (b) What is the probability that at least 8 of them arrive on time?

Practice Problem 10: A restaurant serves 10 different fish entrées, 14 different meat entrées, and 8 different vegetarian entrées. If 5 customers randomly choose their entrées, what is the probability that (a) they all choose vegetarian entrées; (b) at least 3 of them choose vegetarian entrées?

Solutions to Practice Problems

Solution to Practice Problem 1: *What is the probability of getting 2 pairs in a 5-card hand with the other card not matching either pair?*

We are looking for the probability of a hand of the form *AABBC*, where *A*, *B*, and *C* are all different kinds. First, let's find the total number of ways to choose 5 cards from the deck of 52 cards. This is simply the combination $_{52}C_5 = \dfrac{52!}{5!(52-5)!} = 2{,}598{,}960$. This will be the denominator of our probability (the sample space).

Now for the numerator. First, there are $_{13}C_2 = 78$ ways to choose the value of each pair (2, 3, 4, ... , *A*) and there are 2 pairs. Next, there are $_4C_2 = 6$ ways to choose each pair (for example, Jacks and Tens), from each set of 4 in the deck. Next, there are $_{11}C_1 = 11$ ways to choose the remaining card. Finally, there are $_4C_1 = 4$ ways of choosing that card. If we multiply these together, we get $_{13}C_2 \cdot {_4C_2} \cdot {_4C_2} \cdot {_{11}C_1} \cdot {_4C_1} =$ 123,552 total hands that contain a pair.

Thus, the probability of getting 2 pairs is

$$\frac{_{13}C_2 \cdot {_4C_2} \cdot {_4C_2} \cdot {_{11}C_1} \cdot {_4C_1}}{_{52}C_5} = \frac{123{,}552}{2{,}598{,}960} = 0.0475.$$

Solution to Practice Problem 2: *What is the probability of all of the cards in a 5-card hand being of the same suit (a flush)?*

There are 13 cards in any particular suit and we need to choose 5 of them, so there are $_{13}C_5 = 1287$ ways to do so. We need to multiply this number by four because there are four suits. The total number of ways to choose 5 cards from the deck of 52 cards is $_{52}C_5 = \dfrac{52!}{5!(52-5)!} = 2{,}598{,}960$. Therefore, the probability of getting a flush is $\dfrac{_{13}C_5 \cdot 4}{_{52}C_5} = \dfrac{5148}{2{,}598{,}960} = 0.00198$. (For you poker players out there, this includes the probability of a straight flush.)

Solution to Practice Problem 3: *A couple plans to have 8 children. What is the probability that they will have 4 boys and 4 girls?*

First, let's figure out how many different gender sequences are possible. There are 2 possibilities for the child's gender and there are 8 positions in the sequence, so the total number of possible gender sequences is $2^8 = 256$.

Next, let's figure out how many different gender sequences of 4 boys and 4 girls are possible. We could think of this as how many ways can we arrange the letters *GGGGBBBB*. This is a combination of 4 out of a possible 8, or $_8C_4 = 70$.

Thus, the probability of having 4 boys and 4 girls is $\dfrac{_8C_4}{2^8} = \dfrac{35}{128}$.

Solution to Practice Problem 4: *The serial number on a dollar bill consists of 8 digits in a row. What is the probability that the serial number will consist of 5 zeros and 3 ones?*

First, let's figure out how many different serial numbers are possible. There are ten digits and there are eight positions in the serial number, so the total number of possible serial number is $10^8 = 100{,}000{,}000$. Next, let's figure out how many different serial numbers of five zeros and three ones are possible. That is, we want to arrange the digits 0000111. This is a combination of 5 out of a possible 8, or ${}_8C_5 = 56$. Thus, the probability that the serial number will consist of five zeros and three ones is $\frac{{}_8C_5}{10^8} = \frac{56}{100{,}000{,}000}$.

Solution to Practice Problem 5: *A college has 400 freshmen. Two hundred forty of the students will take Calculus, 180 of them will take Writing, and 65 will take both. How many will take neither Calculus nor Writing?*

There are 2 ways to do this. The first is to make a Venn Diagram.

Figure 3

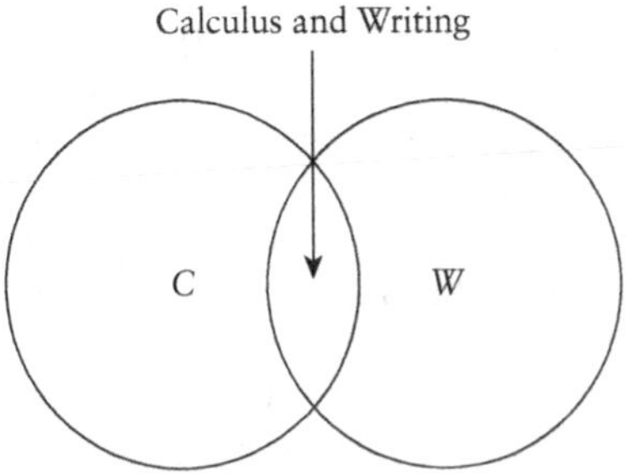

We put 65 in the overlapping region to signify the students who take both Calculus and Writing. Next, we put $240 - 65 = 175$ in the Calculus circle (labeled C), to signify the students who take Calculus but not Writing. Then we put $180 - 65 = 115$ in the Writing circle (labeled W), to signify the students who take Writing but not Calculus. Now we can add these 3 numbers to get $65 + 175 + 115 = 355$. This is the number of students who take Calculus or Writing. Now we subtract this total from the total number of students to get the number of students who take neither Calculus nor Writing: $400 - 355 = 45$.

The other way to get the right answer is to use the formula: A or $B = A + B - (A$ and $B)$. Here we get C or $W = 240 + 180 - 65 = 355$. Then we find $400 - 355 = 45$, which is the number of students who take neither Calculus nor Writing. Of course, it is much faster to use the formula, providing you remember it correctly. Also, if we gave you three groups instead of just two, a Venn Diagram will usually get you to the right answer faster.

Solution to Practice Problem 6: *Suppose that the probability of getting Disease A is 0.12 and the probability of getting Disease B is 0.05. The probability of getting both is 0.02. What is the probability of getting neither Disease A nor Disease B?*

As with the previous problem, there are 2 ways to do this. The first is to make a Venn Diagram.

Figure 4

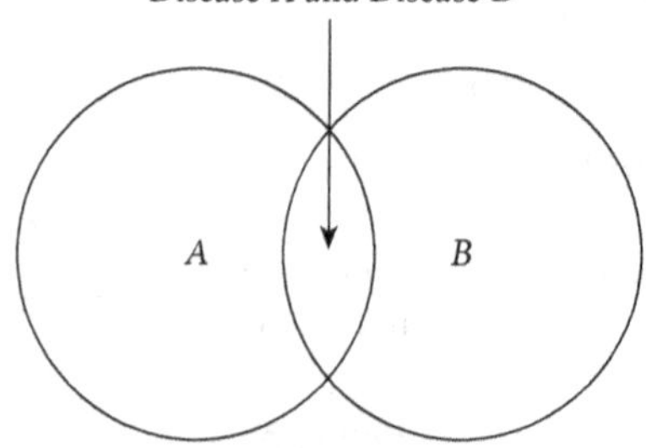

We put 0.02 in the overlapping region to signify the probability of getting both diseases. Next, we put $0.12 - 0.02 = 0.10$ in circle A to signify the probability of getting Disease *A* but not Disease *B*. Then we put $0.05 - 0.02 = 0.03$ in circle *B* to signify the probability of getting Disease *B* but not Disease *A*. We can add these 3 numbers to get $0.10 + 0.03 + 0.02 = 0.15$, which is the probability of getting Disease *A* or Disease *B*. Now we subtract this total from 1 to get the probability of getting neither Disease *A* nor Disease *B*: $1 - 0.15 = 0.85$.

The other way to get the right answer is to use the formula:
$P(A \text{ or } B) = P(A) + P(B) - P(A \text{ and } B)$. Here we get $P(A \text{ or } B) = 0.12 + 0.05 - 0.02 = 0.15$. Then we find $1 - 0.15 = 0.85$, which is the probability of getting neither Disease *A* nor Disease *B*.

Solution to Practice Problem 7: *Eighty men in a high school are asked their heights and if they play basketball. We get the following results.*

	Plays Basketball	**Does Not Play Basketball**
Six Feet Tall and Over	25	8
Under Six Feet Tall	27	20

What is the probability of selecting a man at random and choosing one who (a) does not play basketball; (b) is under six feet tall and plays basketball; (c) is under six feet tall or plays basketball?

(a) There are 8 men who are 6 feet or taller who do not play basketball and 20 men under 6 feet tall who do not play basketball for a total of 28 men who do not play basketball. There 80 men in the survey, so the probability of selecting one who does not play basketball is $\frac{28}{80}$.

(b) There are 27 men who are under 6 feet tall and play basketball, so the probability of selecting one is $\frac{27}{80}$.

(c) There are 47 men who are under 6 feet tall. There are 52 men who play basketball. There are 27 men who are both under 6 feet tall and play basketball. So, if we want to find the number of men who are under 6 feet tall *or* play basketball, we have to be careful not to double count the men who are both under 6 feet tall and play basketball. We can use the formula $P(A \text{ or } B) = P(A) + P(B) - P(A \text{ and } B)$ to get $47 + 52 - 27 = 72$. Thus, the probability of selecting a man who is under 6 feet tall or plays basketball is $\frac{72}{80}$.

Solution to Practice Problem 8: *A test consists of multiple choice questions, each having five possible answers (one of which is correct). If a person guesses on 6 questions, (a) what is the probability that the person gets 3 of them correct? (b) What is the probability that the person gets fewer than 2 questions correct?*

First, let's verify that this is binomial. A person guesses either correctly or incorrectly, the probability is the same for each guess $\left(\frac{1}{5}\right)$ and whether the person guesses correctly on a particular question has nothing to do with whether the person guessed correctly on any previous question. So, this is binomial event.

(a) Now, let's figure out the probability of guessing correctly on 3 of the 6 questions.

The probability that the person guesses correctly on a question is $\frac{1}{5}$, so the probability that the person is incorrect on a question is $1-\frac{1}{5}=\frac{4}{5}$. According to the binomial probability formula, the probability of guessing correctly on 3 of the 6 questions is, thus, $P(3) = {}_6C_3\left(\frac{1}{5}\right)^3\left(\frac{4}{5}\right)^{6-3} = 20\left(\frac{1}{5}\right)^3\left(\frac{4}{5}\right)^3 = \frac{1280}{15{,}625}$.

(b) The probability that the person gets fewer than 2 correct out of 6 questions is the sum of the probabilities of getting 0 and 1 question correct. So, we have to find both of these probabilities and add them. Using the formula, we get:

$$P(0) = {}_6C_0\left(\frac{1}{5}\right)^0\left(\frac{4}{5}\right)^{6-0} = 1\cdot\left(\frac{1}{5}\right)^0\left(\frac{4}{5}\right)^6 = \frac{4096}{15{,}625}$$

$$P(1) = {}_6C_1\left(\frac{1}{5}\right)^1\left(\frac{4}{5}\right)^{6-1} = 6\cdot\left(\frac{1}{5}\right)^1\left(\frac{4}{5}\right)^5 = \frac{6144}{15{,}625}\cdot$$

Thus, the probability of getting fewer than 2 questions correct is $\frac{4096}{15{,}625}+\frac{6144}{15{,}625}=\frac{10{,}240}{15{,}625}$.

Solution to Practice Problem 9: *The probability that an airline flight will arrive on time at a particular airport is 0.767. On a Friday, 10 flights will arrive at the airport. (a) What is the probability that exactly 8 of them will arrive on time? (b) What is the probability that at least 8 of them arrive on time?*

(a) Again, this is a binomial probability question. The probability of a flight being on time is 0.767, so the probability of a flight not being on time is $1-0.767 = 0.233$. Using the formula, we get $P(8) = {}_{10}C_8(0.767)^8(0.233)^{10-8} = 45\cdot(0.767)^8(0.233)^2 = 0.293$.

(b) The probability that at least 8 of them arrive on time is the sum of the probabilities of 8, 9, and all 10 flights being on time. We already have the probability that 8 of them will be on time from part (a), and using the formula, we get:

$$P(9) = {}_{10}C_9(0.767)^9(0.233)^{10-1} = 10\cdot(0.767)^9(0.233)^1 = 0.214$$

$$P(10) = {}_{10}C_{10}(0.767)^{10}(0.233)^{10-10} = 1\cdot(0.767)^{10}(0.233)^0 = 0.070.$$

Now we sum these probabilities to get $0.293 + 0.214 + 0.070 = 0.577$.

Solution to Practice Problem 10: *A restaurant serves 10 different fish entrées, 14 different meat entrées, and 8 different vegetarian entrées. If 5 customers randomly choose their entrées, what is the probability that (a) they all choose vegetarian entrées; (b) at least 3 of them choose vegetarian entrées?*

(a) This may not appear to be binomial because there are 3 different kinds of entrées but it actually is because either a customer chooses a vegetarian entrée or she does not. There are 32 different entrées and 8 of them are vegetarian, so the probability that a customer chooses a vegetarian entrée is $\frac{8}{32}=\frac{1}{4}$. The probability that a customer does not choose a vegetarian entrée is $1-\frac{1}{4}=\frac{3}{4}$. Using the formula, the probability that all 5 of the customers choose a vegetarian entrée is

$$P(5) = {}_5C_5\left(\frac{1}{4}\right)^5\left(\frac{3}{4}\right)^{5-5} = 1\cdot\left(\frac{1}{4}\right)^5\left(\frac{3}{4}\right)^0 = \frac{1}{1024}.$$

(b) The probability that at least 3 of the customers choose vegetarian entrées is the sum of the probabilities that three, four, and all five of them do. We already have the probability of all 5 of them from part (a). Using the formula, we get:

$$P(3) = {}_5C_3\left(\frac{1}{4}\right)^3\left(\frac{3}{4}\right)^{5-3} = 10\cdot\left(\frac{1}{4}\right)^3\left(\frac{3}{4}\right)^2 = \frac{90}{1024}$$

$$P(4) = {}_5C_4\left(\frac{1}{4}\right)^4\left(\frac{3}{4}\right)^{5-4} = 5\cdot\left(\frac{1}{4}\right)^4\left(\frac{3}{4}\right)^1 = \frac{15}{1024}.$$

Thus, the probability that at least 3 of the customers choose vegetarian entrées is $\frac{1}{1024}+\frac{90}{1024}+\frac{15}{1024}=\frac{106}{1024}$.

UNIT FOUR

Conditional Probability

Now we will look at an area of probability that many people find difficult. Sometimes, the probability that an event will occur is affected by something that occurred before it. This previous event can affect the likelihood that the following event occurs. For example, if we draw two cards from a deck, without replacement, the probability that the second card is an Ace depends on the first card.

The *conditional probability* of an event is the probability that an event will occur, given that some prior event has already occurred. The probability that event A will occur, given that event B has already occurred, is written $(A|B)$. This can be found by the quotient of the probability that both events occur and the probability of event B occurring. That is, $P(A|B) = \frac{P(A \text{ and } B)}{P(B)}$. We assume that event B has already occurred and, using that information, calculate the probability that event A will occur. Let's do an example.

Example 1: Suppose we have the following information about a group of 60 people. Some of the people have been inoculated for a disease and some have not, and some have the disease while others do not. (a) If we randomly select 1 of the 60 people, find the probability that the person has the disease given that the person was inoculated. (b) If we randomly select 1 of the 60 people, find the probability that the person was inoculated given that the person has the disease.

	Inoculated	**Not Innoculated**
Disease	8	18
No Disease	23	11

(a) We want to find $P(\text{disease}|\text{inoculated})$. That is, we already know that the person was inoculated, and we want to find the likelihood that the person has the disease. There is a total of 8 + 23 = 31 people who have been inoculated and 8 of them have the disease, so the probability is $\frac{8}{31} \approx 0.258$. If we use the formula, we should get the same result. The probability of having the disease and being inoculated is $P(\text{disease and inoculated}) = \frac{8}{60}$, and the probability of being inoculated is $P(\text{inoculated}) = \frac{31}{60}$. We plug the numbers into the formula: $\frac{\frac{8}{60}}{\frac{31}{60}} = \frac{8}{31}$.

(b) Now we want to find $P(\text{inoculated}|\text{disease})$. Here, we already know that the person has the disease and we want to find the likelihood that the person was inoculated. There is a total of 8 + 18 = 26 people who have the disease, and 8 of them have been inoculated, so the probability is $\frac{8}{26} \approx 0.308$. If we use the formula, we should get the same result. The probability of having the disease and being inoculated is $P(\text{disease and inoculated}) = \frac{8}{60}$, and the probability of having the disease is $P(\text{disease}) = \frac{26}{60}$. We plug these into the formula: $\frac{\frac{8}{60}}{\frac{26}{60}} = \frac{8}{26}$.

Notice that these are not the same probabilities. In the first case, we determined that an inoculated person has an $\frac{8}{31}$ probability of having the disease. In the second case, we determined that a person with the disease has an $\frac{8}{26}$ probability of having been inoculated. Many people confuse the 2 probabilities, which can result in misleading information. Let's do another example.

Example 2: Suppose we have the following information about a group of 100 people who take a test to determine whether they have a serious disease. The test is imperfect and occasionally gives incorrect results. (a) If we randomly select one of the 100 people, find the probability that the person tested positive given that the person has the disease. (b) If we randomly select 1 of the 100 people, find the probability that the person has the disease given that the person tested positive.

	Positive Test	**Negative Test**
Disease	78	7
No Disease	3	12

(a) We want to find $P(\text{positive}|\text{disease})$. There is a total of 78 + 7 = 85 people who have the disease and 78 of them tested positive, so the probability is $\frac{78}{85} \approx 0.918$. If we use the formula, we should get the same result. The probability of having the disease and testing positive is $P(\text{disease and positive}) = \frac{78}{100}$, and the probability of having the disease is $P(\text{disease}) = \frac{85}{100}$. We plug the numbers into the formula: $\frac{\frac{78}{100}}{\frac{85}{100}} = \frac{78}{85}$.

(b) Now we want to find $P(\text{disease}|\text{positive})$. There are a total of 78 + 3 = 81 people who tested positive and 78 of them have the disease, so the probability is $\frac{78}{81} \approx 0.963$. If we use the formula, we should get the same result. The probability of having the disease and testing positive is $P(\text{disease and positive}) = \frac{78}{100}$, and the probability of testing positive is $P(\text{positive}) = \frac{81}{100}$. We plug the numbers into the formula: $\frac{\frac{78}{100}}{\frac{81}{100}} = \frac{78}{81}$.

Sometimes an event does not have an effect on the outcome of another event. If so, the 2 events are called *independent.* If the $P(A|B) = P(A)$, then A and B are *independent* events. Another way to test whether two events are independent is to see if $P(A \text{ and } B) = P(A) \cdot P(B)$. If so, then the 2 events are independent.

Another way to find conditional probability is to use *Bayes' Theorem.* This theorem can be used for some rather complex calculations and we will only touch upon it here. Bayes' Theorem says that $P(A|B) = \frac{P(A) \cdot P(B|A)}{P(A) \cdot P(B|A) + P(not\ A) \cdot P(B|not\ A)}$. Let's do an example.

Example 3: Suppose that there are 2 bakeries that make our famous chocolate chip cookies. We make 70% of our cookies in factory A and 30% of our cookies in factory B. Then, for a randomly selected cookie, the probability that it came from factory A is 0.70. We randomly taste a cookie and it tastes terrible. We know that 4% of the cookies from factory A taste terrible and 3% of the cookies from factory B taste terrible. What is the probability that the terrible-tasting cookie is from factory A?

Let's call T the event where the cookie tastes terrible. Then we want to find $P(A|T)$, the probability that the cookie is from factory A, given that it tastes terrible. Using Bayes' theorem, we will want to evaluate $P(A|T) = \frac{P(A) \cdot P(T|A)}{P(A) \cdot P(T|A) + P(B) \cdot P(T|B)}$. The probability that the cookie is from factory A is $P(A) = 0.70$ and the probability that the cookie is from factory B is $P(B) = 0.30$. We know that 4% of the cookies from factory A taste terrible, so $P(T|A) = 0.04$. We also know that 3% of the cookies from factory B taste terrible, so $P(T|B) = 0.03$. We plug into the formula and we get $P(A|T) = \frac{(0.70)\cdot(0.04)}{(0.70)\cdot(0.04)+(0.30)\cdot(0.03)} = 0.76$. That is, if we taste a randomly selected cookie and it tastes terrible, then the probability that the cookie came from factory A is 0.76.

There is much more to be learned about conditional probability, which is a complex subject and often produces counter-intuitive results. For the purposes of an elementary statistics book, this is all that we will cover. We recommend that you look at a detailed book on probability to learn more about the many aspects of conditional probability and its uses.

Let's practice!

Practice Problems

Practice Problem 1: Suppose we have the following information about a group of 200 students in a statistics class, some of whom studied for the final at least 24 hours in advance and the rest of whom studied less than 24 hours in advance. Some of them passed the final exam while others did not. (a) If we randomly select 1 of the 200 students, find the probability that the student passed the final given that the student studied less than 24 hours in advance. (b) If we randomly select 1 of the 200 students, find the probability that the student studied less than 24 hours in advance given that the student passed the final.

	Studied ≥ 24 hours	**Studied < 24 hours**
Passed the Final	98	85
Did not Pass the Final	7	10

Practice Problem 2: Suppose we have the following information about a group of 480 runners for the local marathon. Some of the runners ran the day before the marathon while others did not. Some of the runners finished the race and some did not. (a) If we randomly select 1 of the 480 runners, find the probability that the runner ran the day before the marathon given that the runner finished the race. (b) If we randomly select 1 of the 480 runners, find the probability that the runner finished the race given that the runner ran the day before the marathon.

	Ran the Day Before	**Did Not Run**
Finished the Marathon	138	301
Did Not Finish the Marathon	32	9

Practice Problem 3: Suppose that we operate 2 factories that make light bulbs. We make 63% of our light bulbs in factory A and 37% of our light bulbs in factory B. We randomly test a light bulb and it is defective. We know that 0.13% of the light bulbs from factory A are defective and 0.24% of the light bulbs from factory B are defective. What is the probability that the defective light bulb is from factory A?

Practice Problem 4: Suppose that there are two high schools in our district and that 55% of the students go to high school W and 45% go to high school M. We also know that on any given day, 6% of the students from high school W are absent and that 11% of the students from high school M are absent. If we randomly choose an absent high school student from the district, what is the probability that the student came from high school W?

Solutions to Practice Problems

Solution to Practice Problem 1: *Suppose we have the following information about a group of 200 students in a statistics class, some of whom studied for the final at least 24 hours in advance and the rest of whom studied less than 24 hours in advance. Some of them passed the final exam while others did not. (a) If we randomly select 1 of the 200 students, find the probability that the student passed the final given that the student studied less than 24 hours in advance. (b) If we randomly select 1 of the 200 students, find the probability that the student studied less than 24 hours in advance given that the student passed the final.*

	Studied ≥ 24 Hours	**Studied < 24 Hours**
Passed the final	98	85
Did not Pass the Final	7	10

(a) We want to find $P(\text{passed}|\text{studied} < 24 \text{ hrs})$. That is, we already know that the student studied less than 24 hours in advance of the final and we want to find the likelihood that the student passed. There is a total of $85 + 10 = 95$ students who studied less than 24 hours in advance, and 85 of them passed, so the probability is $\frac{85}{95} \approx 0.895$. Or we could use the formula. The probability of passing the final and studying less than 24 hours in advance is $P(\text{passed and} < 24 \text{ hrs}) = \frac{85}{200}$, and the probability of studying less than 24 hours in advance is $P(< 24 \text{ hrs}) = \frac{95}{200}$. We plug the numbers into the formula: $\frac{\frac{85}{200}}{\frac{95}{200}} = \frac{85}{95}$.

(b) Now we want to find $P(\text{studied} < 24 \text{ hrs}|\text{passed})$, which reverses the conditions. That is, we already know that the student passed and we want to find the likelihood that the student studied less than 24 hours in advance of the final. There is a total of $85 + 98 = 183$ students who passed, and 85 of them studied less than 24 hours in advance, so the probability is $\frac{85}{183} \approx 0.464$. If we use the formula, we have the probability of passing the final and studying less than 24 hours in advance is $P(\text{passed and} < 24 \text{ hrs}) = \frac{85}{200}$, and the probability of passing the final is $P(\text{passed}) = \frac{183}{200}$. We plug these into the formula: $\frac{\frac{85}{200}}{\frac{183}{200}} = \frac{85}{183}$.

Solution to Practice Problem 2: *Suppose we have the following information about a group of 480 runners for the local marathon. Some of the runners ran the day before the marathon while others did not. Some of the runners finished the race and some did not. (a) If we randomly select 1 of the 480 runners, find the probability that the runner ran the day before the marathon given that the runner finished the race. (b) If we randomly select 1 of the 480 runners, find the probability that the runner finished the race given that the runner ran the day before the marathon.*

	Ran the Day Before	Did Not Run
Finished the Marathon	138	301
Did Not Finish the Marathon	32	9

(a) We want to find $P(\text{ran the day before}|\text{finished})$. There is a total of $138 + 301 = 439$ runners who finished the race, and 138 of them ran the day before, so the probability is $\frac{138}{439} \approx 0.314$. Let's do this with the formula. The probability of finishing the race and running the day before is $P(\text{ran the day before and finished}) = \frac{138}{480}$, and the probability of finishing the race is $P(\text{finished}) = \frac{439}{480}$. We plug the numbers into the formula: $\frac{\frac{138}{480}}{\frac{439}{480}} = \frac{138}{439}$.

(b) Now we want to find $P(\text{finished}|\text{ran the day before})$. There is a total of $138 + 32 = 170$ runners who ran the day before, and 138 of these runners finished, so the probability is $\frac{138}{170} \approx 0.812$. If we use the formula, we have the probability of running the day before and finishing the race is $P(\text{ran the day before and finished}) = \frac{138}{480}$, and the probability of running the day before is $P(\text{ran the day before}) = \frac{170}{480}$. We plug the numbers into the formula: $\dfrac{\frac{138}{480}}{\frac{170}{480}} = \frac{138}{170}$.

Solution to Practice Problem 3: *Suppose that we operate 2 factories that make light bulbs. We make 63% of our light bulbs in factory A and 37% of our light bulbs in factory B. We randomly test a light bulb and it is defective. We know that 0.13% of the light bulbs from factory A are defective and 0.24% of the light bulbs from factory B are defective. What is the probability that the defective light bulb is from factory A?*

Let's call D the event where the light bulb is defective. Then we want to find $P(A|D)$, the probability that the light bulb is from factory A given that it is defective. Using Bayes' theorem, we want to evaluate $P(A|D) = \frac{P(A) \cdot P(D|A)}{P(A) \cdot P(D|A) + P(B) \cdot P(D|B)}$. The probability that the light bulb is from factory A is $P(A) = 0.63$ and the probability that the light bulb is from factory B is $P(B) = 0.37$. We know that 0.13% of the light bulbs from factory A are defective so $P(D|A) = 0.0013$, and we know that 0.24% of the light bulbs from factory B are defective, so $P(D|B) = 0.0024$. We plug the numbers into the formula and we get $P(A|D) = \frac{(0.63) \cdot (0.0013)}{(0.63) \cdot (0.0013) + (0.37) \cdot (0.0024)} = 0.48$. That is, if we test a randomly selected light bulb and it is defective, then the probability that this light bulb came from factory A is 0.48.

Solution to Practice Problem 4: *Suppose that there are two high schools in our district and that 55% of the students go to high school W and 45% go to high school M. We also know that on any given day, 6% of the students from high school W are absent and that 11% of the students from high school M are absent. If we randomly choose an absent high school student from the district, what is the probability that the student came from high school W?*

Let's call A the event that the high school student is absent. Then we want to find $P(W|A)$, the probability that the student is from high school W given that the student is absent. Using Bayes' theorem, we will want to evaluate $P(W|A) = \frac{P(W) \cdot P(A|W)}{P(W) \cdot P(A|W) + P(M) \cdot P(A|M)}$. The probability that a student is from high school W is $P(W) = 0.55$ and the probability that the student is from high school M is $P(M) = 0.45$. We know that 6% of the students from high school W are absent, so $P(A|W) = 0.06$. We know that 11% of the students from high school M are absent, so $P(A|M) = 0.11$. We plug the numbers into the formula and we get $P(W|A) = \frac{(0.55)\cdot(0.06)}{(0.55)\cdot(0.06)+(0.45)\cdot(0.11)} = 0.40$. That is, if we randomly select a student from the district and the student is absent, then the probability that the student came from high school W is 0.40.

UNIT FIVE

Statistics Terms and Experimental Design

Before we delve into our study of Statistics, it is important to learn some terminology and a little about how to conduct an experiment. When we collect observations about an experiment (for example, measurement, the respondent's sex, survey responses) we call this *data* and the methods of planning the experiment, collecting the data, and then presenting, analyzing, and interpreting that data is *Statistics*. The *population* is the complete set of all of the data to be studied and a *sample* is a smaller set of the data, randomly drawn from the population. It is important to remember that the data must be selected in an appropriate fashion, such as randomly. Data that have not been appropriately collected are often useless and any statistical analysis of such data is of no value.

There are 2 main types of data that can be observed–parameters and statistics.

A *parameter* is a measurement of some characteristic of a population. A *statistic* is a measurement of some characteristic of a sample. For example, if 53% of the people in the country voted for Candidate *A* for President, then the number 53% that represents the percentage of people in the country who voted for Candidate *A* is a parameter. On the other hand, if we surveyed 550 people in the country and 53% of those surveyed voted for Candidate *A* for President, then that percentage is a statistic because it is based on a sample, not on the population.

When collecting data, it is usually divided into quantitative and qualitative data. *Quantitative Data* contain measurements or counts. *Qualitative Data* can be separated into categories that cannot be measured numerically. For example, the heights of softball players is quantitative, but their gender is qualitative.

When conducting an experiment, it is of vital importance that the experiment be well designed. One of the most important factors in the design is how to collect the data. Often, when one hears of a statement in the media with a statistic quoted, the statistic is useless because the data were not collected properly. One of the most common and most abused of such statements comes from a *voluntary response sample.* A voluntary response sample is one where the respondent chooses whether to be included in the sample, such as a poll on the internet where a person can choose to respond ("Click here if you like Contestant *A*"). Another example is distributing flyers that ask a person to mail in or call in a response. With such samples, one can only draw valid conclusions about the respondents, not about the population as a whole. We will often find the media generalizing from such polls, but such statements are invalid and can often be misleading.

Another factor to be considered is the size of the sample. If a sample is too small, then we cannot draw valid conclusions about a large population from that sample. Although we will learn how to draw some inferences from small samples, it is important to check that the necessary requirements for such samples be met.

Another problem can be that a sample may appear to be large ("1000 Americans were surveyed"), but valid conclusions can not be drawn about a small subgroup of the population ("kindergarten students from Jefferson County"). We would need to know that the data were collected appropriately and randomly, and that it were large enough to draw conclusions about all of the possible subgroups. This is one of the reasons why election polling is so expensive and difficult, and so often wrong.

Another way that statistical information can be misleading is through the reporting of percentages. Percentages can be used to distort information. For example, suppose that the number of people who are at risk for catching a disease increases from 6 in a million to 9 in a million. This is an absolute increase of 3, but is a 50% increase in the likelihood of catching the disease. Depending on one's bias, either number can be presented, without giving the appropriate context. If the percent of people who participate in a program grows from 40% to 50%, it is a 25% change in the percentage of people participating, not a 10% increase.

When creating a survey, it is important to phrase the question carefully, so that it is not intentionally stated to elicit a particular response. For example, one could ask "Should the President be allowed to refuse to carry out Congress's agenda?" Or, one could ask "Should the President veto the bill?" Also, the order that questions are asked can skew the responses "Do you blame Congress or the President?" versus "Do you blame the President or Congress?" can produce different percentages.

Another way that Statistics are often misused is when correlation is presented as causality. We will explore this later in the book but it is important to realize that *correlation does not imply causality.* For example, we might find that higher family income and test score are correlated, but it does not mean that if you get wealthier, your scores will go up!

When collecting data, it is important to do so in an appropriate way. In an *observational study*, certain characteristics are observed and measured, but there is no attempt to change the characteristic. In an *experiment*, some *treatment* is applied and then its effect is measured. The data may be measured at one point in time, retrospectively, or over a long period of time. In a *cross-sectional study*, the data are collected, observed, and measured at one point in time; in a *retrospective study*, the data are collected from past records; in a *longitudinal study*, the data are collected in the future from different groups that share common factors.

UNIT SIX

Some Basics of Statistics

Now that we have learned some things about Probability, we will shift to Statistics. One of the main differences between the 2 subjects is that probability looks at data and describes what could happen, whereas Statistics looks at data and describes what has already happened. One of the difficulties that students often have with Statistics is that it uses mathematics in unfamiliar ways. There is a lot of new terminology to learn. So, on that note, let's define some terms that we will see in Statistics.

Data are collected observations. They are often in the form of numbers, which can either have their intrinsic value or can symbolize something. For example, the number 10 could either have a value of 10 or could express a 10 on a rating scale, or some other meaning.

Statistics is the set of the methods used to work with the data. Often the data is gathered, then it is organized, presented, analyzed, and interpreted. Finally, one draws conclusions based on the data.

A *population* is the complete set of all of the elements of a subject to be studied. For example, a population could be all of the students in a class, or in a school, or in the country, depending on what is being studied.

A *sample* is a subset of the elements of a population. For example, if a population is all of the students in a school, then a sample might be a selection of 20 students in that school.

Statistics often involves sampling a population. In order for the sample to yield valid results, it is critical that the sample be collected in an appropriate way. Often, such an appropriate way is random selection. Needless to say, this requires careful construction of the experiment and can sometimes be difficult to do. Think of the presidential polls. It is very difficult to draw an accurate conclusion about who is winning from the polls because the sample may not be random, or the error may be too large, or other confounding factors.

There are a variety of ways to take a sample of a population.

One type is a *voluntary response sample*, in which the respondents decide whether or not to be included in the sample. Examples of voluntary response samples are Internet surveys, telephone call-in polls, and so on. Valid conclusions can only be drawn about the people who choose to be included in the sample, but not about the population as a whole. Thus, these types of samples are flawed from a statistical point of view, and we should avoid making statements about a population based on such surveys.

Another type of sample that can be flawed is one in which the sample is too small. Although sometimes we can draw inferences from such samples, they are usually not large enough to enable us to draw conclusions about a population.

Another issue with samples is that the question can be intentionally worded to draw a particular response. For example, consider "Is too much money being wasted on education?" versus "Is too much money being spent on education?"

A good sample is one in which the respondents are chosen randomly, and where the sample is representative of the population as a whole. A random selection process means that each respondent in a sample is equally likely to be chosen. A *simple random sample* is a sample of n subjects where each possible *sample of the same size n* is equally likely to be chosen.

Two other terms that we often see in Statistics are *correlation* and *causality.* Correlation indicates that two variables are related, and that when one changes, the other changes as well. Causality means that one variable causes another variable to change. It is very important to remember: ***Correlation does not imply causality.*** In other words, just because we see a relationship between 2 variables does not mean that one causes the other to respond. For example, greater income is correlated with higher test averages, but this does not mean that having money makes one do better in school or that doing better in school makes one wealthier. We often see stories reported in the media where the story implies a causality that is not necessarily there.

One way to display data is through a *frequency distribution.* This is one that lists data values along with the corresponding frequency of each value.

Example 1: A survey of 100 people measured their total cholesterol levels. The following is a frequency distribution of the results

Cholesterol level	Frequency
Less than 160	3
161–180	7
181–200	11
201–220	22
221–240	51
Greater than 240	6

Note that most of the people have cholesterol levels between 201 and 240. Later, we will learn some tools to help us analyze these results. Assuming that the sample of the 100 people is well chosen, we could then draw some conclusions about the total cholesterol levels of the population as a whole.

It is often convenient to visualize data in the form of a graph or chart. It helps us see patterns that may not be obvious from just looking at the data. One of the most common ways to visualize data is with a *histogram,* which is a bar chart where the horizontal scale represents the classes of data values and the vertical scale represents their frequencies.

Figure 5

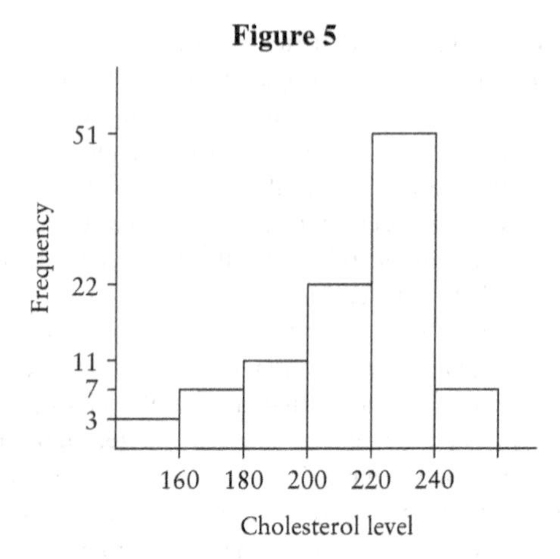

Note that the bars are placed next to each other, without any gaps between them.

A second common type of plot is a *stem-and-leaf plot.* Here the data is represented by separating each value into two parts–the *stem* (for example, the left-most digit, or digits), and the *leaf* (for example, the rightmost digit).

Example 2: Suppose that we have 40 people in our program, with the following ages: 7, 8, 10, 10, 10, 11, 13, 13, 13, 14, 14, 15, 15, 15, 15, 15, 16, 16, 17, 17, 17, 17, 18, 21, 21, 25, 25, 27, 27, 28, 31, 34, 38, 45

Let's construct a stem-and-leaf plot of their ages.

We separate each age into a stem, which will be the tens digit, and a leaf, which will be the units digit. The plot looks like the following:

Stem (tens)	Leaf (units)
0	78
1	00013334455555667778
2	1155778
3	148
4	5

One of the advantages of a stem-and-leaf plot is that we can see the distribution of the frequencies of the data while retaining the original information. Another is that it makes for an easy way to sort the data. The rows of the stem-and-leaf plot are similar to a histogram but on their sides.

Another type of plot is called a *scatter diagram*, or a *scatter plot.* This is useful when we want to look at data where one value is paired to another value. Each pair of values is plotted using the values as x and y coordinates. A dot is placed on the diagram to indicate each pair of values. These kinds of plots are very useful for visualizing correlations between values, which we will explore more in a later unit.

Example 3: We are interested if there is a relationship between a student's Precalculus grade and his or her Physics grade. Here are the grades for 20 students.

Precalculus	75	80	93	87	71	65	68	84	98	77	62	99	79	84	85	99	68	70	79	83
Physics	82	78	86	91	80	72	72	89	95	74	55	94	80	82	80	99	60	62	63	84

Let's make a scatter plot of their grades:

Figure 6

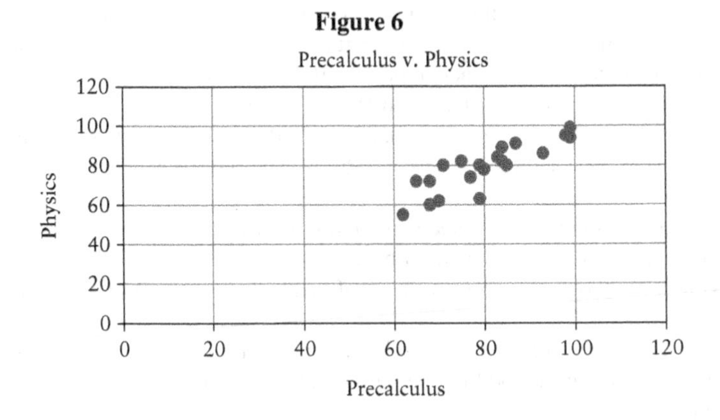

There are of course other ways to visualize data for statistical purposes, but these are the most common. There is one other type of plot, called a *boxplot*, or a *box-and-whisker diagram*, which we will see in a later unit.

Because the purpose of this unit is really just to teach you some common terms and to show you the basic diagrams, we will not be giving you any problems to practice. We suggest that you study this unit carefully to make sure that you are comfortable with all of the material presented here.

UNIT SEVEN

Center and Spread

Now that we have learned how to display and to think about data, it is time to learn how to use Statistics to analyze data. One of the first things that is useful to know is the average of a set of numbers. Why is this useful? Well, suppose we know that the average summer daytime temperature in your town for a given week is 84°. If one day's temperature is 90 degrees, we know that it is hotter than it usually is. One thing that we will learn in Statistics is how unusual it is for the temperature to be 90°. For example, suppose the temperatures for that week, in degrees, had been {83, 84, 85, 84, 83, 85, 84}. Then it looks as if 90° is unusual. Suppose instead that the temperatures, in degrees, had been {89, 91, 90, 84, 78, 77, 79}. Now 90° looks much more likely. By the way, the collection of values is called a *data set*, and each number in the set is called a *data point*. Let's examine the concept of average in more detail.

In Statistics, the word *average* is not really very helpful because there is more than one type of average. When most people use the word *average*, they are usually referring to the *arithmetic mean* of a set of data, which we will refer to simply as the *mean*.

The *mean* of a set of data is found by summing the values of the *data points* in the set and dividing the total by the number of data points in the set. Let's do an example.

Example 1: Find the means of the two data sets {83, 84, 85, 84, 83, 85, 84} and {89, 91, 90, 84, 78, 77, 79}.

We simply add up the 7 data points in each set and divide by 7. For the first set, we get $\frac{83+84+85+84+83+85+84}{7}=\frac{588}{7}=84$. For the second set, we get $\frac{89+91+90+84+78+77+79}{7}=\frac{588}{7}=84$. Thus, each set of data has a mean of 84.

Another type of average for a set of data is the *median*. The *median* of a set of numbers is the center of a set of data when the numbers are arranged in increasing or decreasing order of magnitude.

Example 2: Find the median of the set {83, 84, 85, 84, 83, 85, 84}.

In order to find the median, we first put the numbers in order: {83, 83, 84, 84, 84, 85, 85}. The median of the numbers is 84.

There are some subtleties to median that we need to address. Note that in the above set, there are 7 data points, so the center of the set will be the fourth value when they are placed in order. There are then 3 values greater than or equal to the median, and 3 values less than or equal to the median. Note that there can be values equal to the median.

Whenever there is an odd number of data points in a set, the median will be a value in the data set because we can divide the set at that middle value and there will be the same number of values above and below the median. For example, if there are 21 data points in a set, then the median will be the eleventh value because there will be 10 values above the median and 10 values below it.

But what about when there is an even number of data points in a set? Now we have a small problem. For example, if there are 10 data points and we split the set down the middle, then there will be 2 sets of 5 values but no middle value. What do we do? We average the 2 numbers closest to the center of the set (after we have put the values in order). Let's do an example.

Example 3: Find the median of {82, 80, 85, 84, 88, 81, 90, 76}.

First, we put the data in order: {76, 80, 81, 82, 84, 85, 88, 90}. Note that there are 8 data points, so we will average the 2 values closest to the center: $\frac{82+84}{2}=83$. Thus, the median of the set is 83. Note that there are 4 values less than 83 and 4 values greater than 83.

Also, note that if there is an odd number of data points in a set, the median will be a number in that set, whereas if there is an even number of data points in a set, the median will not necessarily be a number in the set. In Example 3, the median, 83, is not in the set.

Example 4: Find the median of {84, 80, 85, 84, 88, 81, 90, 76}.

First, we put the data in order: {76, 80, 81, 84, 84, 85, 88, 90}. Now we average the 2 values closest to the center: $\frac{84+84}{2}=84$, so the median of the set is 84. Note that this time the median is a member of the set.

A third type of average is called the *mode*. The *mode* of a set of numbers is the value that occurs most frequently. When 2 values occur with the same greatest frequency, they are both modes and the set is called *bimodal*. When more than 2 values occur with the same greatest frequency, they are all modes and the set is called *multimodal*. When no value occurs more than once, the set has no mode.

Example 5: Find the mode(s) of each of the following sets: (a) {83, 84, 85, 84, 83, 85, 84}; (b) {81, 84, 85, 84, 83, 85, 82}; (c) {83, 84, 85, 84, 81, 85, 83}; (d) {83, 84, 85, 81, 86, 82, 80}.

(a) We look at the values in the set and we see that 83 and 85 each occur twice, and 84 occurs 3 times, so the mode is 84.

(b) We look at the values in the set and we see that 81, 82, and 83 each occur once, and 84 and 85 each occur twice. So this set is bimodal, and the modes are 84 and 85.

(c) We look at the values in the set and we see that 81 occurs once, and 83, 84, and 85 each occur twice. So this set is multimodal, and the modes are 83, 84 and 85.
(d) We look at the values in the set and we see that each value occurs once with no value occurring more frequently than any other value. This set has no mode.

Each of these averages tells us something different about a data set, and each is useful in a different way. For example, suppose we know that the mean weight of a student in a school is 150 pounds. If the school's elevator can hold 1500 pounds, then the elevator can hold an average of 10 students. One problem with a mean is when a data set is distorted by an overly large or small value (called an *outlier*). For example, let's say that we have 5 people and 4 of them have an income of $50,000 a year, but 1 of them has an income of $1,000,000 a year. The mean income is $240,000 (add the incomes and divide by 5). If we said that the average income of the group of people was $240,000 we would be giving the impression that each person has an income of around that number. But this is not really true. How could we solve this problem? One solution is to throw out the outlier and recalculate the mean. Now we would get $50,000, which seems better. Another option is to use the median instead. The median of the group is also $50,000, which gives us a better indication of people's incomes.

Of course, there are times that a median is not that useful either. Suppose we have a test that is scored from 0 to 100 and the median score is 90. We know that half of the students scored at or above 90, and half scored at or below 90. But, we do not know much about the individual scores. They could have been {88, 89, 90, 90, 90, 91, 92}, which means that everyone scored right around 90. Or they could have been {48, 70, 72, 90, 92, 95, 100}. These scores are much more widely spread. Also, notice that the top score is only 10 points above the median, but the bottom score is 42 points below. In a little bit, we will learn some useful tools to tell us more about how data are spread out.

When is the mode useful? Well, suppose you owned a shoe store. You want to know which size shoe sells the most, not the average shoe size or the middle shoe size.

These 3 averages are measures of the center of a distribution and combined can tell us a lot about a set of data. One thing we can learn is how much the data is *skewed.* A set of data is *skewed* if it is not distributed symmetrically and extends more to one side than the other. A *symmetric* data distribution means that the data on the left side of the center and the data on the right side of the center approximately mirror each other.

Suppose we displayed the distribution of a data set graphically. If a histogram looks something like this, it is roughly symmetric.

Figure 7

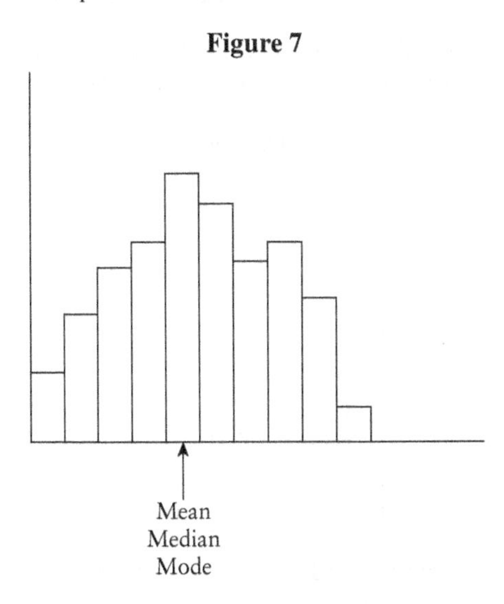

In a symmetric distribution, the mean, median, and mode are approximately the same.

If our histogram looks like the figure below, it is *skewed left* (sometimes called *negatively skewed*). This means that it has a long tail to the left side, illustrating that the numbers less than the center tend to be spread out widely, whereas the numbers greater than the center tend to be more tightly bunched.

Figure 8

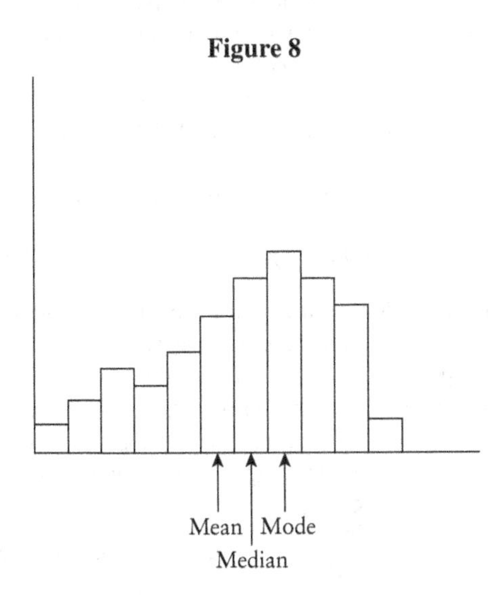

In a skewed left distribution, the mean and median are to the *left* of the mode.

If our histogram looks like the figure below, it is *skewed right* (sometimes called *positively skewed*). This means that it has a long tail to the right side.

Figure 9

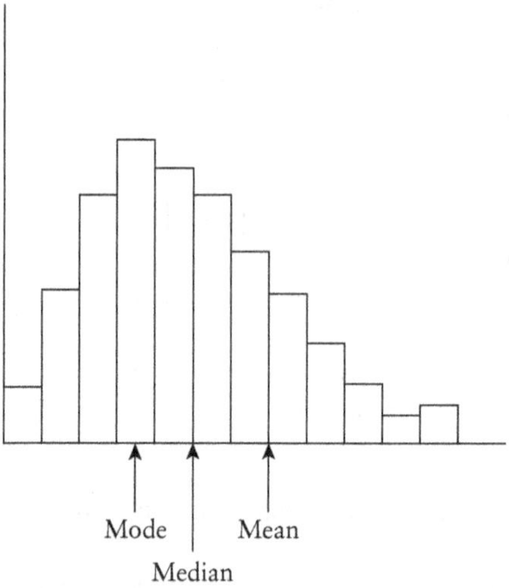

Mode
Median
Mean

In a skewed right distribution, the mean and median are to the *right* of the mode.

Now, let's take a moment to learn a little notation:

We will generally use x or x_i to stand for an individual value in a set of data.

We will use $\overline{x}$ to stand for the mean of a set of sample values.

We will use μ to stand for the mean of all the values in a population.

We will use Σ to stand for summing a set of values.

For example, we find the mean of a set of data by the following formula: $\overline{x} = \frac{\sum_{i=1}^{n} x_i}{n}$. The x_i stands for the individual values in the set. The variable n tells us how many values are in the set. The $\sum_{i=1}^{n} x_i$ tells us to start with the first value, then add the second value, then add the third, and so on, until we add the nth (last) value. Finally, we divide the total by the number of values.

There is one last type of mean to look at, called a *weighted mean* or a *weighted average*. We find this type of mean when we want to assign different weights to different values in a data set. We multiply each of the values by its weight to get its *weighted value*. Then we sum the weighted values and divide by the sum of the weights. The formula is: $\overline{x} = \frac{\sum_{i=1}^{n} x_i * w_i}{\sum_{i=1}^{n} w_i}$, where w_i stands for the weight of each value. Let's do an example.

Example 6: Suppose your math grade is calculated in the following way. Homework is 10% of your grade, quizzes are 20%, the two midterms are 15% each, and the final exam is 40%. You have a score of 91 on your homework, 88 on your quizzes, 94 on the first midterm, 86 on the second midterm, and 91 on the final. What is your math grade?

We take each value and multiply it by its weight. Homework is $(91)\left(\frac{10}{100}\right)=9.1$, quizzes are $(88)\left(\frac{20}{100}\right)=17.6$, the first midterm is $(94)\left(\frac{15}{100}\right)=14.1$, the second midterm is $(86)\left(\frac{15}{100}\right)=12.9$, and the final is $(91)\left(\frac{40}{100}\right)=36.4$. Now we add these weighted values to get 9.1 + 17.6 + 14.1 + 12.9 + 36.4 = 90.1. We divide this number by the total of the weights: $\frac{10}{100}+\frac{20}{100}+\frac{15}{100}+\frac{15}{100}+\frac{40}{100}=1$. Thus, your math grade is $90.1/1=90.1$. By the way, is this a good grade? It depends on how the rest of the class did!

The way that the distribution of values in a data set is spread out tells us about the *variation* of the data. One useful measure is the *range* of the data. The *range* of a set of data is simply the highest value in the set minus the lowest value in the set. For example, let's look at the data sets we used when discussing median. In the set {88, 89, 90, 90, 90, 91, 92}, the range of values is 92 – 88 = 4. In the set {48, 70, 72, 90, 92, 95, 100}, the range is 100 – 48 = 52. As we can see, the range of values in the second set is much greater than the range of values in the first set. This is useful to know but it does not tell us a lot about the data set. Now let's learn a much more valuable tool to look at the variation of a set of data.

The *standard deviation* of a set of data is a measure of variation of values about the mean. It is a type of average deviation of values about the mean. The formula for the standard deviation of a population of data is found by: $\sigma=\sqrt{\frac{\sum_{i=1}^{n}(x_i-\mu)^2}{N}}$.

Let's translate this formula. In order to find the standard deviation of a set of data, we first find the mean (μ). Next we take each value in the set and subtract the mean from it ($x_i-\mu$), to find the deviation of each value from the mean. We square each of these deviations and sum them. Then we divide by N, the number of values in the population, which gives us an average of the squares of the deviations. Finally, we take the square root of this number to get the standard deviation.

Why do we square the deviations? Suppose we had the following set of number: {–10, 0, 10}. The mean of these numbers is 0, and we have 3 deviations: –10 – 0 = –10, 0, and 10 – 0 = 10. Now let's average those 3 deviations. We get 0. But clearly we have values that deviate from the mean. The problem is that the deviations, –10 and 10, cancelled each other. We want to get rid of the sign of each deviation and just use the magnitude of it. We can do this by squaring each deviation, to make each one positive. When we are done, we just have to remember to take the square root so we get the same units that we started with.

Let's do an example.

Example 7: Suppose we have the weights of a group of 10 students: {112, 123, 145, 100, 109, 116, 112, 119, 94, 122}. Find the mean and the standard deviation of the students' weights.

First, let's find the mean. We sum the students' weights and divide by 10. We get

$$\bar{\mu} = \frac{112+123+145+100+109+116+112+119+94+122}{10} = \frac{1152}{10} = 115.2$$

Now for the fun part. Let's find the standard deviation. It's often useful to make a table to organize the calculations.

x_i	$x_i - \mu$	$(x_i - \mu)^2$
112	112 – 115.2 = –3.2	10.24
123	123 – 115.2 = 7.8	60.84
145	145 – 115.2 = 29.8	888.04
100	100 – 115.2 = –15.2	231.04
109	109 – 115.2 = –6.2	38.44
116	116 – 115.2 = 0.8	0.64
112	112 – 115.2 = –3.2	10.24
119	119 – 115.2 = 3.8	14.44
94	94 – 115.2 = –21.2	449.44
122	122 – 115.2 = 6.8	46.24
	$\sum_{i=1}^{10}(x_i - \mu)^2 =$	1749.6

Now that we have found the sum of the deviations, we simply divide by 10 to get their average, $\frac{1749.6}{10} = 174.96$. We then take the square root: $\sqrt{174.96} \approx 13.23$.

This was not as hard as you might have thought, but it can be quite tedious, especially if you have a large data set. Fortunately, most scientific calculators can calculate a standard deviation. Also, programs like EXCEL, and, of course, dedicated statistics programs, can calculate a standard deviation. Of course, one still has to enter the data, but you can't have everything!

In the next unit, we will learn more about what to do with a standard deviation once we have found it. For now, we will just calculate it.

This standard deviation is for a set of data that represents the whole population that one is looking at. For example, suppose one wanted to find the standard deviation of the weights of the junior class of a high school. If the class has hundreds of students, this might be a cumbersome and time-consuming task. Instead, one could take a representative sample of the students and find the standard deviation of the sample instead. In later units, we will look at sampling in more depth. For now, let's learn how to calculate a sample standard deviation. There is a slight difference in notation. For a population standard deviation, we use the lower case Greek letter *sigma* (σ), and we use the lower case Greek letter *mu* (μ) for the population mean. For a sample standard deviation, we use the lower case letter s, and we use the $\bar{x}$ ("*x-bar*") for the sample mean. The formula is the same except that we divide the sum of the deviations by $n - 1$ (where n is the number of values in the sample). We subtract 1 from the number of values. We do this for reasons that are a bit complex for this book but suffice it to say that if we divided by n, we would consistently underestimate the value of the standard deviation.

Here are a few important properties of the standard deviation:

- The standard deviation is always either positive or zero (which only happens when all of the data values are the same).
- A larger standard deviation indicates a greater amount of variation of values in the data set.
- Outliers can greatly increase the value of the standard deviation.
- The units of standard deviation are the same as for the data values.

Example 8: Suppose we were manufacturing boxes of cereal and we want to make sure that the weight of the boxes is consistent. We manufacture thousands of boxes a day and decide to take a sample of 12 boxes. We get the following weights (in ounces): {20.1, 19.8, 19.9, 20.0, 20.2, 20.15, 19.8, 20.1, 20.05, 19.9, 20.75, 20}. Find s, the sample standard deviation.

First, let's find the mean. We sum the students' weights and divide by 12. We get

$$\bar{x} = \frac{20.1+19.8+19.9+20.0+20.2+20.15+19.8+20.1+20.05+19.9+20.75+20.0}{12}$$

$$= \frac{240.75}{12} = 20.0625.$$

Once again, let's make a table of our calculations:

x_i	$x_i - \bar{x}$	$(x_i - \bar{x})^2$
20.1	20.1 − 20.0625 = 0.0375	0.00140625
19.8	19.8 − 20.0625 = −0.2625	0.06890625
19.9	19.9 − 20.0625 = −0.1625	0.02640625
20.0	20.0 − 20.0625 = −0.0625	0.00390625
20.2	20.2 − 20.0625 = 0.1375	0.01890625
20.15	20.15 − 20.0625 = 0.0875	0.00765625
19.8	19.8 − 20.0625 = −0.2625	0.06890625

20.1	20.1 − 20.0625 = 0.0375	0.00140625
20.05	20.05 − 20.0625 = −0.0125	0.00015625
19.9	19.9 − 20.0625 = −0.1625	0.02640625
20.75	20.75 − 20.0625 = 0.6875	0.47265625
20.0	20.0 − 20.0625 = −0.0625	0.00390625
	$\sum_{i=1}^{12}(x_i - \overline{x})^2 =$	0.700625

Now that we have found the sum of the deviations, we simply divide by 11 to get the mean. (Remember that when we are finding the standard deviation of a sample, we divide by 1 less than that number of values in the sample.). We get $\frac{0.700625}{11} = 0.0636931818$. We then take the square root: $\sqrt{0.0636931818} \approx 0.2524$.

Thus, our cereal boxes have an average weight of 20.0625 ounces, and the average deviation from that weight is approximately 0.2524 ounces. We will learn more in the next unit about how to interpret this information.

One last measure of variation is the *variance*. The variance of a data set is the square of the standard deviation. The variance is important in certain types of statistical analyses, which we will see later in this book. Note that the units of variance are different than the units of the original data set. In the previous example, the variance would be 0.06369 square ounces. The unit "square ounces" has no real meaning, so for now variance is not very useful and we will concentrate on standard deviation.

Time to practice!

Practice Problems

For Practice Problems 1–5, find the (a) mean, (b) median, (c) mode, and (d) range of the following data sets.

Practice Problem 1: The heights of all 12 freshmen in a Statistics seminar are: {66, 61, 63, 68, 66, 69, 62, 63, 62, 66, 63, 63}.

Practice Problem 2: The following are the numbers of bottles of water sold in a 2-week period at the student union snack bar: {214, 278, 232, 305, 310, 269, 297, 305, 215, 300, 317, 293, 214, 323}.

Practice Problem 3: The following are the numbers of right answers on a 20-question math test given to a class of 11 students: {16, 8, 20, 19, 18, 17, 16, 17, 19, 20, 17}.

Practice Problem 4: A college has 4000 students. We wish to examine the number of hours in a week that a typical student spends studying. We polled 20 of them and received the following results: {20, 6, 32, 14, 11, 30, 19, 17, 25, 26, 28, 26, 32, 21, 32, 18, 26, 18, 45, 20}.

Practice Problem 5: A high school has 2000 students. We wish to examine the Body Mass Index of a typical student. We measured 20 of them and received the following results: $\begin{Bmatrix} 19.7, 23.6, 19.7, 22.9, 29.2, 25.2, 24.1, 21.4, 22.0, 25.2, \\ 34.1, 20.7, 36.3, 29.6, 25.8, 35.1, 25.2, 28.5, 18.2, 40.1 \end{Bmatrix}$.

Practice Problem 6: Find the standard deviation of the freshmen heights in Practice Problem 1.

Practice Problem 7: Find the standard deviation of the number of bottles of water in Practice Problem 2.

Practice Problem 8: Find the standard deviation of the numbers of right answers in Practice Problem 3.

Practice Problem 9: Find the standard deviation of the number of hours of studying in Practice Problem 4.

Practice Problem 10: Find the standard deviation of the Body Mass Indices in Practice Problem 5.

Solutions to Practice Problems

For Practice Problems 1–5, find the (a) mean, (b) median, (c) mode, and (d) range of the following data sets.

Solution to Practice Problem 1: *The heights of all 12 freshmen in a Statistics seminar are: {66, 61, 63, 68, 66, 69, 62, 63, 62, 66, 63, 63}.*

(a) We find the mean by adding up the numbers and dividing by 12. We get

$$\overline{x} = \frac{66+61+63+68+66+69+62+63+62+69+65+63}{12} = \frac{777}{12} = 64.75.$$

(b) In order to find the median, we first put the numbers in order of increasing value. We get {61, 62, 62, 63, 63, 63, 65, 66, 66, 68, 69, 69}. There are 12 numbers, so in order to find the median, we take the 2 numbers closest to the middle and average them. We get $\frac{63+65}{2} = 64$.

(c) The mode is the number that occurs most frequently. Now that we have put the numbers in order, we can see that the mode is 63.

Solution to Practice Problem 2: *The following are the numbers of bottles of water sold in a 2-week period at the student union snack bar: {214, 278, 232, 305, 310, 269, 297, 305, 215, 300, 317, 293, 214, 323}.*

(a) We find the mean by adding up the numbers and dividing by 14. We get

$$\mu = \frac{214+278+232+305+310+269+297+305+215+300+317+293+214+329}{14}$$

$$= \frac{3878}{14} = 277.$$

(b) In order to find the median, we first put the numbers in order of increasing value. We get {214, 214, 215, 232, 269, 278, 293, 297, 300, 305, 305, 310, 317, 329}. There are 14 numbers, so in order to find the median, we take the 2 numbers closest to the middle and average them. We get $= \frac{293+297}{2} = 295$.

(c) The mode is the number that occurs most frequently. Now that we have put the numbers in order, we can see that there are two modes: 214 and 305 (the distribution is bimodal).

Solution to Practice Problem 3: *The following are the numbers of right answers on a 20-question math test given to a class of 11 students: {16, 8, 20, 19, 18, 17, 16, 17, 19, 20, 17}.*

(a) We find the mean by adding up the numbers and dividing by 11. We get

$$\mu = \frac{16+8+20+19+18+17+16+17+19+20+17}{11} = \frac{187}{11} = 17.$$

(b) In order to find the median, we first put the numbers in order of increasing value. We get {8, 16, 16, 17, 17, 17, 18, 19, 19, 20, 20}. There are 11 numbers, so the median will be the number in the middle, 17.

(c) The mode is the number that occurs most frequently. Now that we have put the numbers in order, we can see that the mode is 17.

Solution to Practice Problem 4: *A college has 4000 students. We wish to examine the number of hours in a week that a typical student spends studying. We polled 20 of them and received the following results: {20, 6, 32, 14, 11, 30, 19, 17, 25, 26, 28, 26, 32, 21, 32, 18, 26, 18, 45, 20}.*

Note that here we are sampling from a population. This will affect how we calculate the standard deviation, and we use some different notation.

(a) We find the mean by adding up the numbers and dividing by 20. We get

$$\bar{x} = \frac{20+6+32+14+11+30+19+17+25+26+28+26+32+21+32+18+26+18+45+20}{20}$$

$$= \frac{466}{20} = 23.3.$$

(b) In order to find the median, we first put the numbers in order of increasing value. We get {6, 11, 14, 17, 18, 18, 19, 20, 20, 21, 25, 26, 26, 26, 28, 30, 32, 32, 32, 45}. There are 20 numbers, so in order to find the median, we take the 2 numbers closest to the middle and average them. We get $\frac{21+25}{2} = 23$.

(c) The mode is the number that occurs most frequently. Now that we have put the numbers in order, we can see that there are two modes, 26 and 32. The distribution is bimodal.

Solution to Practice Problem 5: *A high school has 2000 students. We wish to examine the Body Mass Index of a typical student. We measured twenty of them and received the following results:*

$$\begin{Bmatrix} 19.7, 23.6, 19.7, 22.9, 29.2, 25.2, 24.1, 21.4, 22.0, 25.2, \\ 34.1, 20.7, 36.3, 29.6, 25.8, 35.1, 25.2, 28.5, 18.2, 40.1 \end{Bmatrix}.$$

(a) We find the mean by adding up the numbers and dividing by 20. We get

$$\bar{x} = \frac{19.7+23.6+19.7+22.9+29.2+25.2+24.1+21.4+22.0+25.2+34.1+20.7+36.3+29.6+25.8+35.1+25.2+28.5+18.2+40.1}{20}$$

$$= \frac{526.6}{20} = 26.33.$$

(b) In order to find the median, we first put the numbers in order of increasing value. We get {18.2, 19.7, 19.7, 20.7, 21.4, 22, 22.9, 23.6, 24.1, 25.2, 25.2, 25.2, 25.8, 28.5, 29.2, 29.6, 34.1, 35.1, 36.3, 40.1}. There are twenty numbers, so in order to find the median, we take the 2 numbers closest to the middle and average them. We get $\frac{25.2+25.2}{2} = 25.2$.

(c) The mode is the number that occurs most frequently. Now that we have put the numbers in order, we can see that the mode is 25.2.

Solution to Practice Problem 6: *Find the standard deviation of the freshmen heights in Practice Problem 1.*

We found the mean of the data set in Practice Problem 1 ($\mu = 64.75$). Now, let's make a table of our calculations:

x_i	$x_i - \mu$	$(x_i - \mu)^2$
66	66 – 64.75 = 1.25	1.5625
61	61 – 64.75 = –3.75	14.0625
63	63 – 64.75 = –1.75	3.0625
68	68 – 64.75 = 3.25	10.5625
66	66 – 64.75 = 1.25	1.5625
69	69 – 64.75 = 4.25	18.0625
62	62 – 64.75 = –2.75	7.5625
63	63 – 64.75 = –1.75	3.0625
62	62 – 64.75 = –2.75	7.5625

69	69 – 64.75 = 4.25	18.0625
65	65 – 64.75 = 0.25	0.0625
63	63 – 64.75 = –1.75	3.0625
	$\sum_{i=1}^{12}(x_i-\mu)^2 =$	88.25

Now that we have found the sum of the deviations, we simply divide by 12 to get the mean. We get $\frac{88.25}{12}=7.35416\overline{6}$ and take the square root: $\sigma=\sqrt{7.35416\overline{6}}\approx 2.7119$.

Solution to Practice Problem 7: *Find the standard deviation of the number of bottles of water in Practice Problem 2.*

We found the mean of the data set in Practice Problem 2 ($\mu = 277$). Now, let's make a table of our calculations:

x_i	$x_i-\mu$	$(x_i-\mu)^2$
214	214 – 277 = –63	3969
278	278 – 277 = 1	1
232	232 – 277 = –45	2025
305	305 – 277 = 28	784
310	310 – 277 = 33	1089
269	269 – 277 = –8	64
297	297 – 277 = 20	400
305	305 – 277 = 28	784
215	215 – 277 = –62	3844
300	300 – 277 = 23	529
317	317 – 277 = 40	1600
293	293 – 277 = 16	256
214	214 – 277 = –63	3969
329	329 – 277 = 52	2704
	$\sum_{i=1}^{14}(x_i-\mu)^2 =$	22,018

Now that we have found the sum of the deviations, we simply divide by 14 to get the mean. We get $\frac{22{,}018}{14} = 1572.714286$ and then take the square root: $\sigma = \sqrt{1572.714286} \approx 39.6575$.

Solution to Practice Problem 8: *Find the standard deviation of the numbers of right answers in Practice Problem 3.*

We found the mean of the data set in Practice Problem 3($\bar{\mu}$ = 17). Now, let's make a table of our calculations:

x_i	$x_i - \bar{\mu}$	$(x_i - \bar{\mu})^2$
16	$16 - 17 = -1$	1
8	$8 - 17 = -9$	81
20	$20 - 17 = 3$	9
19	$19 - 17 = 2$	4
18	$18 - 17 = 1$	1
17	$17 - 17 = 0$	0
16	$16 - 17 = -1$	1
17	$17 - 17 = 0$	0
19	$19 - 17 = 2$	4
20	$20 - 17 = 3$	9
17	$17 - 17 = 0$	0
	$\sum_{i=1}^{11} (x_i - \bar{\mu})^2 =$	110

Now that we have found the sum of the deviations, we simply divide by 11 to get the mean. We get $\frac{110}{11} = 10$ and then take the square root: $\sigma = \sqrt{10} \approx 3.1623$.

Solution to Practice Problem 9: *Find the standard deviation of the number of hours of studying in Practice Problem 4.*

We found the mean of the data set in Practice Problem 4($\bar{x} = 23.3$). Now, let's make a table of our calculations:

x_i	$x_i - \bar{x}$	$(x_i - \bar{x})^2$
20	$20 - 23.3 = -3.3$	10.89
6	$6 - 23.3 = -17.3$	299.29

32	32 – 23.3 = 8.7	75.69
14	14 – 23.3 = –9.3	86.49
11	11 – 23.3 =–12.3	151.29
30	30 – 23.3 = 6.7	44.89
19	19 – 23.3 = –4.3	18.49
17	17 – 23.3 = –6.3	39.69
25	25 – 23.3 = 1.7	2.89
26	26 – 23.3 = 2.7	7.29
28	28 – 23.3 = 4.7	22.09
26	26 – 23.3 = 2.7	7.29
32	32 – 23.3 = 8.7	75.69
21	21 – 23.3 = –2.3	5.29
32	32 – 23.3 = 8.7	75.69
18	18 – 23.3 = –5.3	28.09
26	26 – 23.3 = 2.7	7.29
18	18 – 23.3 = –5.3	28.09
45	45 – 23.3 = 21.7	470.89
20	20 – 23.3 = –3.3	10.89
	$\sum_{i=1}^{20}(x_i-\mu)^2 =$	1468.2

Now that we have found the sum of the deviations, we simply divide by 19 to get the mean. Note that we do not divide by 20, the total number of values in the sample. As we discussed previously, when finding a sample standard deviation, we divide by 1 less than the total number. We get $\frac{1468.2}{19} = 77.27368421$ and then take the square root: $s = \sqrt{77.27368421} \approx 8.7905$.

Solution to Practice Problem 10: *Find the standard deviation of the Body Mass Indices in Practice Problem 5.*

We found the mean of the data set in Practice Problem 5 ($\bar{x} = 26.33$). Now, let's make a table of our calculations:

x_i	$x_i - \bar{x}$	$(x_i - \bar{x})^2$
19.7	$19.7 - 26.33 = -6.63$	43.9569
23.6	$23.6 - 26.33 = -2.73$	7.4529
19.7	$19.7 - 26.33 = -6.63$	43.9569
22.9	$22.9 - 26.33 = -3.43$	11.7649
29.2	$29.2 - 26.33 = 2.87$	8.2369
25.2	$25.2 - 26.33 = -1.13$	1.2769
24.1	$24.1 - 26.33 = -2.23$	4.9729
21.4	$21.4 - 26.33 = -4.93$	24.3049
22	$22 - 26.33 = -4.33$	18.7489
25.2	$25.2 - 26.33 = -1.13$	1.2769
34.1	$34.1 - 26.33 = 7.77$	60.3729
20.7	$20.7 - 26.33 = -5.63$	31.6969
36.3	$36.3 - 26.33 = 9.97$	99.4009
29.6	$29.6 - 26.33 = 3.27$	10.6929
25.8	$25.8 - 26.33 = -0.53$	0.2809
35.1	$35.1 - 26.33 = 8.77$	76.9129
25.2	$25.2 - 26.33 = -1.13$	1.2769
28.5	$28.5 - 26.33 = 2.17$	4.7089
18.2	$18.2 - 26.33 = -8.13$	66.0969
40.1	$40.1 - 26.33 = 13.77$	189.6129
	$\sum_{i=1}^{20}(x_i - \bar{\mu})^2 =$	707.002

Now that we have found the sum of the deviations, we simply divide by 19 to get the mean. Note that we do not divide by 20, the total number of values in the sample. As we discussed previously, when finding a sample standard deviation, we divide by 1 less than the total number. We get $\frac{707.002}{19} = 37.21063158$ and then take the square root: $s = \sqrt{37.21063158} \approx 6.100$.

UNIT EIGHT

Exploring Data and Introducing the *z* Score

Now that we have learned how to find a mean, median, mode, and standard deviation, let's learn how to analyze data using these measures. We will often encounter data of probabilities that are randomly distributed in a population. Such data will usually have a *bell-shaped* distribution. That is, when we graph the data, the curve will look like a bell. The curve looks like this:

Figure 10

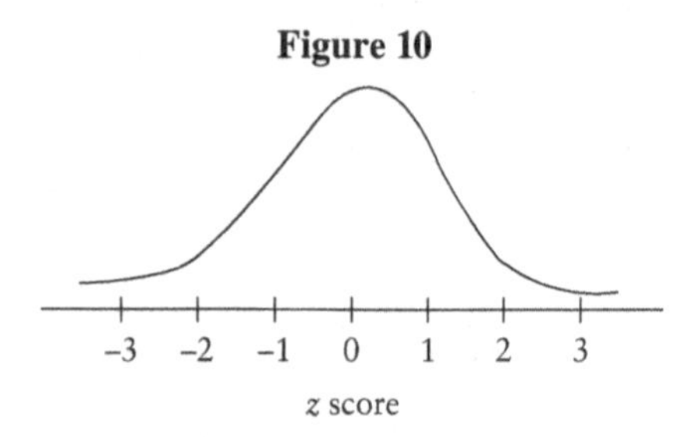

If we are looking at something that occurs randomly, we will usually get a symmetrical distribution like the one above. When this happens, there are many types of analyses that we can do on such distributions.

First, let's look at what is called the *Empirical Formula.* This is very handy when interpreting values for a standard deviation. The rule applies to data sets that have a distribution that is approximately bell-shaped. If so:

- Approximately 68% of all values fall within 1 standard deviation of the mean.
- Approximately 95% of all values fall within 2 standard deviations of the mean.
- Approximately 99.7% of all values fall within 3 standard deviations of the mean.

Note that 99.7% is practically all of the values. This is another way of saying that it is unusual to encounter a value that is more than 3 standard deviations above or below the mean.

Example 1: IQ scores of adults on the Wechsler IQ test (there are many kinds of IQ tests) have a bell-shaped distribution with a mean of 100 and a standard deviation of 15. What is the approximate percentage of adults have scores between 70 and 130?

The standard deviation of IQ scores is 15, so two standard deviations is 30. If we take the mean and subtract two standard deviations (2 standard deviations below the mean), we get a score of $100 - 2(15) = 70$. Similarly, if we take the mean and add 2 standard deviations (2 standard deviations above the mean), we get a score of 100 + 2(15) = 130. Thus, according to the Empirical Formula, we expect approximately 95% of the IQ scores to be between 70 and 130.

If we look a little more closely at the bell curve, we can use its symmetry to get a little more information about standard deviations.

If 68% of values are within one standard deviation of the mean, then approximately 34% are within 1 standard deviation below the mean and approximately 34% are within 1 standard deviation above the mean. Similarly, because 95% – 68% = 27% and half of 27% is 13.5%, we can find how many scores are between 1 and 2 standard deviations of the mean. We can make a nice diagram of the different intervals around the mean:

Figure 11

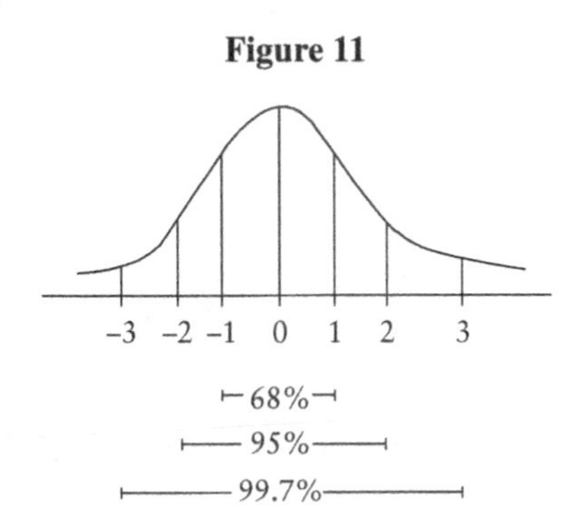

Now, we can use this information to answer slightly more difficult problems.

Example 2: Given the IQ information in Example 1, approximately what percentage of adults have scores between 85 and 130?

Because the standard deviation is 15 and the mean is 100, 1 standard deviation below the mean is a score of 85, and 2 standard deviations above the mean is 130. Note that approximately 34% lie within 1 standard deviation below the mean, and approximately 34% + 13.5% = 47.5% lie with 2 standard deviations above the mean. If we add the two percentages, we get 34% + 47.5% = 81.5%.

At this point we might ask: what if what we are looking for is not an integral multiple of the standard deviation? Also, we would like our answers to be a little more precise. We can find these answers using a *z score.* A z score is found by converting a value to a standardized scale. We will use z scores a lot in the remainder of this book, so this is very important to learn.

A *z score*, or *standard score*, is the number of standardized deviations that a given value x is above or below the mean, using the following equation:

$z = \dfrac{x-\mu}{\sigma}$ (for populations) or $z = \dfrac{x-\overline{x}}{s}$ (for samples). We will usually round z scores to 2 decimal places.

Example 3: What is the z score for a person whose IQ is 125? In other words, how many standard deviations above the mean is the person's IQ?

The IQ information is for the population at large, so we use the z score equation $z = \frac{125 - 100}{15} = 1.67$. That is, the person's IQ is 1.67 standard deviations above the mean. We will learn a little later in the book how to figure out how unusual such an IQ is.

Example 4: The mean height of a man in the United States is 69.1 inches with a standard deviation of 2.82 inches. The mean height of a woman in the United States is 63.6 inches with a standard deviation of 2.51 inches. Is it more unusual for a man to be 76 inches tall (6 ft. 4 in.) or for a woman to be 72 inches tall (6 ft.)?

Let's compute the z score for each.

Man: $z = \frac{76 - 69.1}{2.82} = 2.45$

Woman: $z = \frac{72 - 63.6}{2.51} = 3.35$

The man is 2.45 standard deviations above the mean and the woman is 3.35 standard deviations above the mean, so it is more unusual to meet a woman who is 6 feet tall than a man who is 6 feet, four inches tall.

Example 5: How tall would the man in the previous example have to be to be 3.35 standard deviations above the mean?

Now we just solve the z score equation for x: $z = \frac{x - 69.1}{2.82} = 3.35$. We get $x = (3.35)(2.82) + 69.1 = 78.55$ inches (a little over six feet, six inches).

We will learn much more about interpreting z scores but for now keep in mind that a z score greater than 2 or less than –2 is "unusual." (By the way, a positive z score means that a value is greater than the mean and a negative score means that a value is less than the mean.) If we use heights as an example, this would mean that it would be unusual to see a man who is taller than 74.74 inches tall (close to six feet, three inches) or shorter than 63.46 inches (close to five feet, three inches). Ask yourself how often you see a man who is either above or below those heights. Your answer should be: "Not very often." This is what we mean by *unusual*. We do not mean that it never happens, but that we do not expect it to happen very often. The 2-standard-deviation interval corresponds to 95% of values and is a measure of "unusualness" that is fairly standard in Statistics. We will see this more in later units. When researchers want to refer to something that is very unusual, they will often use a shorthand expression like "more than three Sigmas" to refer to something that is more than 3 standard deviations from the mean, which is a very unusual occurrence. After all, how many men do you see who are six feet, nine inches tall?

Another common way to divide up data is to use *percentiles* and *quartiles*.

A sorted set of data can be divided into 99 percentiles, each of which refers to the percentage of values less than that percentile. For example, if a value is in the 80th percentile then 80 percent of values are less than that value. The highest possible percentile is thus the 99th percentile, not the 100th. We often see percentiles used for test results, as a way of comparing how someone did versus the rest of the test takers.

A sorted set of data can be divided into four quartiles:

- The first quartile (Q_1) corresponds to the lowest 25% of the values. (That is, at least 75% of the values are greater than or equal to Q_1 and 25% of the values are less than or equal to Q_1.)
- The second quartile (Q_2) corresponds to the bottom (or top) 50% of the values. Note that the second quartile is the same as the median (M).
- The third quartile (Q_3) corresponds to the highest 25% of the values. (That is, at least 25% of the values are greater than or equal to Q_3 and 75% of the values are less than or equal to Q_3.)

Example 6: Group the following data set into quartiles: {8, 15, 22, 11, 6, 1, 15, 30, 17, 9, 2, 14}.

First, we need to sort the data set: {1, 2, 6, 8, 9, 11, 14, 15, 15, 17, 22, 30}. (By the way, we could sort the data in either ascending or descending order. It is simply customary to sort from lowest to highest.) There are 12 values, so the quartiles will be in groups of 3. Notice that the 3 lowest values are {1, 2, 6}. Just as we did with the median, we now find Q_1 by averaging the third lowest and the fourth lowest values: $\frac{6+8}{2} = 7$. Thus, $Q_1 = 7$. In other words, 25% of the values are below 7, and 75% of the values are above 7.

Next, we find Q_2 in the same way as we did with the median. We average the 2 values closest to the middle of the data set: $\frac{11+14}{2} = 12.5$. Thus, $Q_2 = 12.5$. Now let's find Q_3. We use the same procedure as for the other 2 quartiles. We get $\frac{15+17}{2} = 16$. Thus, $Q_3 = 16$, which means that 25% of the values are above 16 and 75% are below 16.

Sometimes it is convenient to give a *five-number summary* of a data set. This consists of the minimum value, Q_1, the median (or Q_2), Q_3, and the maximum value. Another term that we will encounter is the *interquartile range* (IQR). This is simply $Q_3 - Q_1$. It stands for the middle 50% of a data set.

One way to display a data set is to make a *boxplot* (or *box-and-whisker diagram*), which consists of a line extending from the minimum value to the maximum value, and a box with lines drawn at Q_1, Q_2, and Q_3. This enables one to easily visualize how dispersed a data set is, the interquartile range, the value of the median, and so on.

Example 7: Display the data from Example 6 in a boxplot.

We have already found the relevant values: $Q_1 = 7$, $Q_2 = 12.5$, and $Q_3 = 16$. The lowest value is 1 and the highest value is 30. The boxplot looks like this:

Figure 12

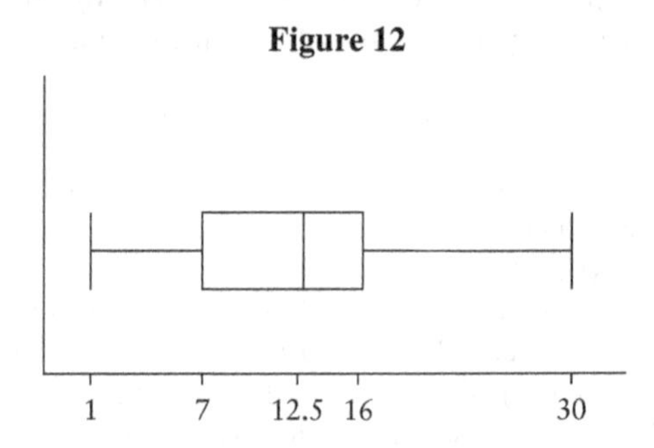

Note how far the maximum value line is from the box relative to the minimum value line. It is customary to define an outlier as a value that is either 1.5 *IQR* below Q_1 or 1.5 *IQR* above Q_3. In Example 7, the $IQR = 16 - 7 = 9$, and $(1.5)(9) = 13.5$. Thus, an outlier is a value that is above $16 + 13.5 = 29.5$ or below $7 - 13.5 = -6.5$. There is one value, 30, that is an outlier.

Time to practice!

Practice Problems

Practice Problem 1: Suppose that the typical 18-year-old man consumes 2300 calories a day, with a standard deviation of 400 calories. Use the Empirical Formula to estimate what percentage of 18-year-old men consume between 1500 and 3100 calories a day?

Practice Problem 2: Suppose that the average age of an American car is now 11.1 years, with a standard deviation of 3.2 years. Use the Empirical Formula to estimate what percentage of cars are of an age between (a) 14.3 years and 7.9 years; (b) greater than 14.3 years?

Practice Problem 3: Given the calorie information in Practice Problem 1, approximately what percentage of 18-year-old men consume between 1100 and 2700 calories a day?

Practice Problem 4: Given the age information in Practice Problem 2, approximately what percentage cars are of an age between 11.1 and 17.5 years?

Practice Problem 5: Given the calorie information in Practice Problem 1, what is the z score for an 18-year-old man who consumes 2550 calories a day?

Practice Problem 6: Given the age information in Practice Problem 2, what is the z score of a car that is 18.2 years old?

Practice Problem 7: In Town A, in a typical 30-day stretch, the mean number of days of rain is 11 with a standard deviation of 4. In Town B, in a typical 30-day stretch, the mean number of days of rain is 22, with a standard deviation of 3. Is it more unusual if Town A has a 30-day stretch with no rain or if Town B has a 30-day stretch where it rains every day?

Practice Problem 8: A typical fish of species X lays a mean number of 3000 eggs with a standard deviation of 220 eggs. A typical fish of species Y lays a mean number of 5000 eggs with a standard deviation of 550 eggs. Is it more unusual for a fish of species X to lay 2300 eggs or for a fish of species Y to lay 7000 eggs?

Practice Problem 9: In Practice Problem 7, how many days in a 30 day stretch would it have to rain in Town A for the number of days to be 2.5 standard deviations above the mean?

Practice Problem 10: In Practice Problem 8, how many eggs would fish Y have to lay to be 1.7 standard deviations below the mean?

Practice Problem 11: Group the following PSAT scores into quartiles. Give the five-number summary of the data:

$$\begin{Bmatrix} 64, 48, 55, 58, 66, 72, 70, 59, 77, 39 \\ 59, 55, 56, 73, 71, 55, 30, 79, 62, 50 \end{Bmatrix}.$$

Practice Problem 12: Group the following set of test scores into quartiles. Give the five-number summary of the data:

$$\begin{Bmatrix} 100, 92, 87, 58, 93, 100, 79, 89, 100, 60, 77, 76 \\ 59, 83, 88, 90, 100, 44, 74, 65, 86, 66, 79, 75, 82 \end{Bmatrix}.$$

Practice Problem 13: Make a box plot of the data in Practice Problem 11.

Practice Problem 14: Make a box plot of the data in Practice Problem 12.

Solutions to Practice Problems

Solution to Practice Problem 1: *Suppose that the typical 18-year-old man consumes 2300 calories a day, with a standard deviation of 400 calories. Use the Empirical Formula to estimate what percentage of 18-year-old men consume between 1500 and 3100 calories a day?*

The standard deviation of calories is 400, so 2 standard deviations is 800. If we take the mean and subtract 2 standard deviations, we get $2300 - 2(400) = 1500$ calories.

Similarly, if we take the mean and add 2 standard deviations, we get 2300 + 2(400) = 3100 calories. Thus, according to the Empirical Formula, we expect approximately 95% of the 18-year-old men to consume between 1500 and 3100 calories a day.

Solution to Practice Problem 2: *Suppose that the average age of an American car is now 11.1 years, with a standard deviation of 3.2 years. Use the Empirical Formula to estimate what percentage of cars are of an age between (a) 14.3 years and 7.9 years; (b) greater than 14.3 years?*

(a) The standard deviation of a car's age is 3.2 years. If we take the mean and subtract 1 standard deviation, we get 11.1 – 3.2 = 7.9 years. Similarly, if we take the mean and add 1 standard deviation, we get 11.1 + 3.2 = 14.3 years. Thus, according to the Empirical Formula, we expect approximately 68% of the American cars to be between 14.3 and 7.9 years old.

(b) If 68% of American cars are between 14.3 and 7.9 years old, then 100% – 68% = 32% of cars are either greater than 14.3 years old or less than 7.9 years old. The distribution of car ages is approximately symmetrical (without evidence to the contrary), so there should be an equal percent of cars greater than 14.3 years old, or less than 7.9 years old. This means that we should expect approximately 16% of the cars to be greater than 14.3 years old.

Solution to Practice Problem 3: *Given the calorie information in Practice Problem 1, approximately what percentage of 18-year-old men consume between 1100 and 2700 calories a day?*

Because the standard deviation is 400 and the mean is 2300, 3 standard deviations below the mean is a score of 2300 – 3(400) = 1100, and 1 standard deviations above the mean is 2300 + 400 = 2700. Note that approximately 34% + 13.5% + 2.4% = 49.9% lie within 3 standard deviations below the mean, and approximately 34% lie within 1 standard deviation above the mean. If we add the 2 percentages, we get 49.9% + 34% = 83.9%.

Solution to Practice Problem 4: *Given the age information in Practice Problem 2, approximately what percentage cars are of an age between 11.1 and 17.5 years?*

Because the standard deviation is 3.2 and the mean is 11.1, 2 standard deviations above the mean is a score of 11.1 + 2(3.2) = 17.5. Note that approximately 34% + 13.5% = 47.5% lie within two standard deviations above the mean.

Solution to Practice Problem 5: *Given the calorie information in Practice Problem 1, what is the z score for an 18-year-old man who consumes 2550 calories a day?*

The calorie information is for the population at large, so we use the z score equation $z = \frac{2550 - 2300}{400} = 0.625$. That is, a person who consumes 2550 calories a day is 0.625 standard deviations above the mean.

Solution to Practice Problem 6: *Given the age information in Practice Problem 2, what is the z score of a car that is 18.2 years old?*

We use the z score equation $z = \frac{18.2 - 11.1}{3.2} = 2.22$. That is, a car that is 18.2 years old is 2.22 standard deviations above the mean.

Solution to Practice Problem 7: *In Town A, in a typical 30 day stretch, the mean number of days of rain is 11 with a standard deviation of 4. In Town B, in a typical 30 day stretch, the mean number of days of rain is 22, with a standard deviation of 3. Is it more unusual if Town A has a 30 day stretch with no rain, or if Town B has a 30 day stretch where it rains every day?*

Let's compute the z score for each.

Town A: $z = \frac{0 - 11}{4} = -2.75$

Town B: $z = \frac{30 - 22}{3} = 2.67$

Town A is 2.75 standard deviations below the mean and Town B is 2.67 standard deviations above the mean, so it is more unusual for Town A to have 30 days without rain than for Town B to have 30 days of rain.

Solution to Practice Problem 8: *A typical fish of species X lays a mean number of 3000 eggs with a standard deviation of 220 eggs. A typical fish of species Y lays a mean number of 5000 eggs with a standard deviation of 550 eggs. Is it more unusual for a fish of species X to lay 2300 eggs or for a fish of species Y to lay 7000 eggs?*

Let's compute the z score for each.

Species X: $z = \frac{2300 - 3000}{220} = -3.18$

Species Y: $z = \frac{7000 - 5000}{550} = 3.64$

Species X is 3.18 standard deviations below the mean and Species Y is 3.64 standard deviations above the mean, so it is more unusual for Species Y to lay 7000 eggs than for Species X to lay 2300 eggs.

Solution to Practice Problem 9: *In Practice Problem 7, how many days in a 30 day stretch would it have to rain in Town A for the number of days to be 2.5 standard deviations above the mean?*

Now we just solve the z score equation for x: $z = \frac{x - 11}{4} = 2.5$. We get $x = (2.5)(4) + 11 = 21$.

Solution to Practice Problem 10: *In Practice Problem 8, how many eggs would fish Y have to lay to be 1.7 standard deviations below the mean?*

Now we just solve the z score equation for x: $z = \frac{x - 5000}{550} = -1.7$. We get $x = (-1.7)(550) + 5000 = 4065$.

Solution to Practice Problem 11: *Group the following PSAT scores into quartiles. Give the five-number summary of the data:*

$$\begin{Bmatrix} 64, 48, 55, 58, 66, 72, 70, 59, 77, 39 \\ 59, 55, 56, 73, 71, 55, 30, 79, 62, 50 \end{Bmatrix}.$$

First, we need to sort the data set: {30, 39, 48, 50, 55, 55, 55, 56, 58, 59, 59, 62, 64, 66, 70, 71, 72, 73, 77, 79} . (By the way, we could sort the data in either ascending or descending order. It is simply customary to sort from lowest to highest.). There are 20 values, so the quartiles will be in groups of 5. The 5 lowest values are {30, 39, 48, 50, 55}. Just as we did with the median, we now find Q_1 by averaging the fifth lowest and the sixth lowest values: $\frac{55+55}{2} = 55$. Thus, $Q_1 = 55$. In other words, 25% of the values are below 55, and 75% of the values are above 55.

Next, we find Q_2 in the same way as we did with the median. We average the 2 values closest to the middle of the data set: $\frac{59+59}{2} = 59$. Thus, $Q_2 = 59$. Now let's find Q_3. We use the same procedure as for the other 2 quartiles. We get $\frac{70+71}{2} = 70.5$. Thus, $Q_3 = 70.5$, which means that 25% of the values are above 70.5, and 75% are below 70.5.

The five-number summary is thus: *Minimum* = 30, $Q_1 = 55$, *Median* (Q_2) = 59, $Q_3 = 70.5$, and *Maximum* = 79.

Solution to Practice Problem 12: *Group the following set of test scores into quartiles. Give the five- number summary of the data:*

$$\begin{Bmatrix} 100, 92, 87, 58, 93, 100, 79, 89, 100, 60, 77, 76 \\ 59, 83, 88, 90, 100, 44, 74, 65, 86, 66, 79, 75, 82 \end{Bmatrix}.$$

First, we need to sort the data set: {44, 58, 59, 60, 65, 66, 74, 75, 76, 77, 79, 79, 82, 83, 86, 87, 88, 89, 90, 92, 93, 100, 100, 100, 100}. There are 25 values, so the quartiles will be in 4 groups of 6 with the median alone in the middle. The 6 lowest values are {44, 58, 59, 60, 65, 66}. Just as we did with the median, we now find Q_1 by average the sixth lowest and the seventh lowest values: $\frac{66+74}{2} = 70$. Thus, $Q_1 = 70$. In other words, 25% of the values are below 70, and 75% of the values are above 70.

Next, we find Q_2 the same way that we did the median. The middle number is 82. Thus, $Q_2 = 82$. Now let's find Q_3. We use the same procedure as for the other two quartiles. We get $\frac{90+92}{2} = 91$. Thus $Q_3 = 91$, which means that 25% of the values are above 91 and 75% are below 91.

The five-number summary is thus: *Minimum* = 44, $Q_1 = 70$, *Median* (Q_2) = 82, $Q_3 = 91$, and *Maximum* = 100.

Solution to Practice Problem 13: *Make a box plot of the data in Practice Problem 11.*

We have already found the relevant values: $Q_1 = 55$, $Q_2 = 59$ and $Q_3 = 70.5$. The lowest value is 30, and the highest value is 79. The boxplot looks like this:

Figure 13

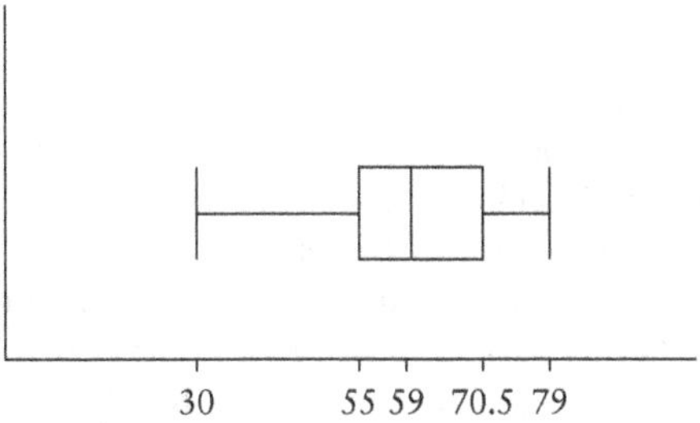

30 55 59 70.5 79

Solution to Practice Problem 14: *Make a box plot of the data in Practice Problem 12.*

We have already found the relevant values: $Q_1 = 70$, $Q_2 = 82$ and $Q_3 = 91$. The lowest value is 44, and the highest value is 100. The boxplot looks like this:

Figure 14

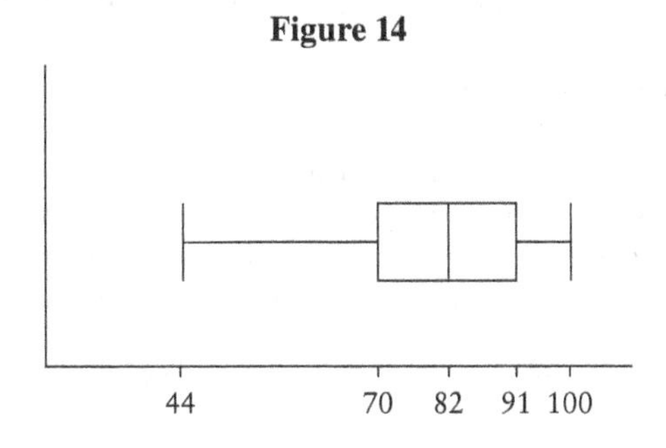

UNIT NINE

Probability Distributions and More About the z Test

In this unit, we will combine some of what we learned in the Probability section of this book with some of what we have learned about Statistics. Once again, we will begin by learning a few terms. First, we will be working with *random variables*. A *random variable* is a variable that has a particular value that is determined by chance each time we follow a procedure. We will usually represent a random variable by x. A *probability distribution* is a distribution of the probabilities of each possible value of a random variable. Let's do an example.

Example 1: In City A, the probability that it rains on any given day is 0.25. Find the probability distribution for the possible number of days of rain in one week.

We can find the probability that it rains a given number of days in a week using the binomial probability formula that we learned in Unit Three. Recall that the probability that a binomial event will occur r times out of a possible n is $P(r) = {}_nC_r\,(p^r)\,(q^{n-r})$, where:

n is the total number of events;

r is the number of desired events;

p is the probability that the desired event occurs;

q is the probability that the desired event does *not* occur (that is, $1 - p$).

Here, $n = 7$, $p = 0.25$, $q = 1 - 0.25 = 0.75$, and r varies from 0 to 7.

Let's find all of the probabilities. The probabilities that it rains r days of the week are:

$$P(0) = {}_7C_0(0.25^0)(0.75^7) = 0.1335$$

$$P(1) = {}_7C_1(0.25^1)(0.75^6) = 0.3115$$

$$P(2) = {}_7C_2(0.25^2)(0.75^5) = 0.3115$$

$$P(3) = {}_7C_3(0.25^3)(0.75^4) = 0.1730$$

$$P(4) = {}_7C_4(0.25^4)(0.75^3) = 0.0577$$

$$P(5) = {}_7C_5(0.25^5)(0.75^2) = 0.0115$$

$$P(6) = {}_7C_6(0.25^6)(0.75^1) = 0.0013$$

$$P(7) = {}_7C_7(0.25^7)(0.75^0) = 0.000061.$$

The probabilities form a probability distribution. Sometimes, it can be convenient to represent them in a table or a histogram. For example, here is a table of the probabilities:

r	$P(r)$
0	0.01335
1	0.3115
2	0.3115
3	0.173
4	0.0577
5	0.0115
6	0.0013
7	0.000061

Notice some things about our results. First, given that the probability that it rains on a particular day is 0.25 and that (0.25)(7) = 1.75, we expect that, most weeks, it should rain somewhere between 1 and 2 days a week. In our probability distribution, the most likely numbers of days of rain are either 1 or 2 days. Second, if we add up the probabilities, the total is 1 (allowing for rounding). This leads us to a couple of rules.

In a probability distribution:

- For each value of x, $0 \le P(x) \le 1$.
- For all possible values of x, $\sum_{i=0}^{n} P(x_i) = 1$.

In other words, first, an individual probability is between 0 and 1, inclusive. This should make sense. Try to think of either a negative probability or a probability of something happening being greater than 100%. Second, the sum of all of the probabilities is 1.

By the way, the random variable that we are working with here is called a *discrete random variable*. A *discrete random variable* is a variable that has a countable number of values or a finite number of values. For example, we can count the number of days in a week where the number of days is 0 through 7, inclusive. Also, we cannot have something like 3.2 days in a week.

A *continuous random variable* is one that can have infinitely many values, and any value is possible. For example, the amount of rain that we can get could be any value from 0 to (hypothetically) infinity. We could get 1 inch, or 1.01 inches, or 1.001 inches, or any other value.

In Example 1, we estimated that we should get around 1.75 days of rain a week. We can make a better guess by calculating the *expected value* of the random variable. More precisely, the *expected value* of a random value is the average value of the outcomes, and is found by $E = \sum_{i=1}^{n} [x_i \cdot P(x_i)]$.

Example 2: Calculate the expected value for the number of days of rain in Example 1.

Let's use the formula for expected value. We need to multiply each outcome by the probability of the outcome and sum them. We get

$$E = (0)(0.01335) + (1)(0.3115) + (2)(0.3115) + (3)(0.173) + (4)(0.0577) + (5)(0.0115) + (6)(0.0013) + (7)(0.000061) = 1.750027.$$

Thus the expected value for the number of days of rain in a week is 1.750027. Our initial guess was pretty good!

Let's do another example.

Example 3: A simple gambling game has the following rules. You place a bet for \$1 on whether you can guess the winning number, which is a number from 00 to 99. If you are correct, you will win \$75. What is the expected value of your winnings?

There are two possible probabilities–either you will guess the correct number or you will not. There are 100 numbers, so the probability that you will guess correctly is $\frac{1}{100}$. Thus, the probability that you will not guess correctly is $1-\frac{1}{100}=\frac{99}{100}$. If you guess correctly, you would win \$75. You had to bet \$1, so your winnings will be \$75 – \$1 = \$74. If you guess incorrectly, you would win –\$1 (that it, you will lose \$1). If we take each outcome, multiply it by its probability and sum them, we will get the expected value: $(74)\left(\frac{1}{100}\right)+(-1)\left(\frac{99}{100}\right)=-\frac{25}{100}=-0.25$. Your expected value is that you will lose \$0.25. Not much of a game is it? By the way, this is a simplified version of the old "numbers" games that many criminal organizations used to run and is also a simplified version of the lottery. We will leave you to draw your own conclusions about playing the lottery.

The probability distribution that we looked at in Example 1 is called a *binomial probability distribution* because the outcomes are binomial; that is, either it rains or it does not. For binomial probability distributions, we can find the mean and standard deviation using the following formulas:

$$\mu = np$$

$$\sigma = \sqrt{npq},$$

where p is the probability that the outcome occurs and q is the probability that it does not occur.

Example 4: Suppose we examine the number of boys born in 20 births. Suppose that the probability that a boy is born in any particular birth is 0.50. Find the mean and standard deviation of the number of boys born in 20 births.

We simply use the formulas above, where $n = 20$, $p = 0.50$, and $q = 1 - 0.50 = 0.50$. We get $\mu = (20)(0.50) = 10$ and $\sigma = \sqrt{(20)(0.50)(0.50)} \approx 2.236$. This means that we expect, on average, that 10 boys will be born out of every 20 births (not a surprise). We can use the Empirical Formula to say that 95% of the time, the number of boys should be between $10-(2)(2.236)=5.528$ and $10+(2)(2.236)=14.472$.

So far, we have looked at discrete probability distributions. Now let's look at a probability where the variable is a continuous random variable. Remember in the last unit we looked at a distribution that is symmetric and bell-shaped. Technically, the formula for such a distribution is $y=\frac{e^{-\frac{(x-\mu)^2}{2\sigma^2}}}{\sigma\sqrt{2\pi}}$. Do not worry about the formula. It is not something that we will address in this book. If a continuous random variable has such a distribution, we call this a *normal distribution.* A *probability density function* is a graph of a continuous probability distribution where the total area under the curve is 1 and where every point on the curve is on or above the x-axis. Finally, a *standard normal distribution* is a probability density function where the mean is 0, the standard deviation is 1 and the area under the density curve is 1.

If we have a standard normal distribution, we can find the probability of an outcome using the z score that we learned in the previous unit. The z score will give the area under the curve from negative infinity up to a vertical line above a specific value of z. We can find these areas using a table (see Table 1 in the back of this book), a calculator equipped to find such scores, EXCEL, or a statistical software package, such as Minitab or STATDISK. We read the table by first going down the leftmost column until we get to the units and tenths digits of the number we are looking for. Then we read across until we get to the column that corresponds to the hundredths digit of what we are looking for. This is the probability of a value less than or equal to our desired z score. Let's do an example.

Example 5: Suppose that the length of a football game is normally distributed with an average length of 200 minutes and a standard deviation of 15 minutes. What is the probability that a football game is less than 180 minutes?

First, we compute the z score. We get $z=\frac{180-200}{15}=-1.33$. Let's make a sketch to see what this looks like on a bell curve:

Figure 15

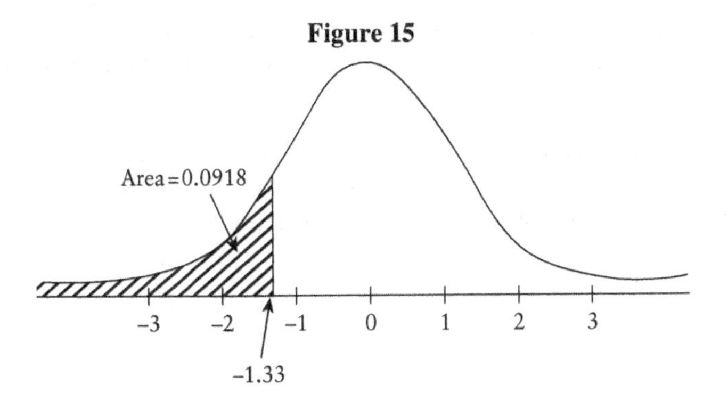

Now we go to the z table in the book and we go down the left column until we get to –1.3. Now we go across to the column under the heading 0.03, which gives us the z score for –1.33, namely $P(z \leq -1.33) = 0.0918$. In other words, the probability that a football game lasts less than 180 minutes is 0.0918.

Example 6: Suppose that the math scores on the SAT are normally distributed with an average of 500, and a standard deviation of 70. What is the probability that a math score on the SAT will be greater than 700?

First, we compute the z score. We get $z = \dfrac{700 - 500}{70} = 2.86$. Let's make a sketch to see what this looks like on a bell curve:

Figure 16

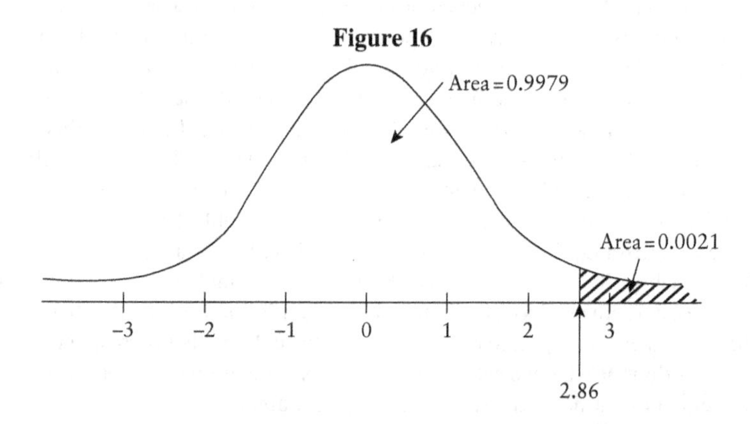

Now we go to the z table in the book and we go down the left column until we get to 2.8. Now we go across to the column under the heading 0.06, which gives us the z score for 2.86, namely $P(z \leq 2.86) = 0.9979$. In other words, the probability that a SAT math score is less than 700 is 0.9979. Of course, this means that the probability that a SAT math score is greater than 700 is $1 - 0.9979 = 0.0021$.

The z table will give us the probability that a z score is less than a particular number. And, as we have just read, we can find the probability that a z score is greater than a particular number by finding the value from the z table and subtracting it from 1.

Example 7: The lifetime of a particular type of light bulb is normally distributed, with a mean of 2000 hours with a standard deviation of 350 hours. What is the probability that a light bulb will last for more than 2500 hours?

First, we compute the z score. We get $z = \frac{2500 - 2000}{350} = 1.43$. Let's make a sketch to see what this looks like on a bell curve:

Figure 17

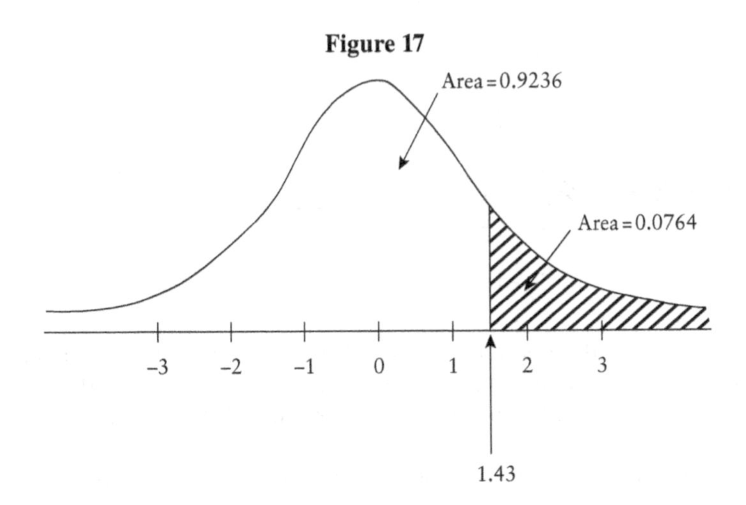

Now we go to the z table in the book and we go down the left column until we get to 1.4. Now we go across to the column under the heading 0.03, which gives us the z score for 1.43, namely $P(z \leq 1.43) = 0.9236$. In other words, the probability that a light bulb lasts for less than 2500 hours is 0.9236. Now we can find the probability that the light bulb will last more than 2500 hours, which is $1 - 0.9236 = 0.0764$.

We can also find the probability that we will get between 2 values using z scores. If we want to find $P(a < z < b)$, we find the probability that the value is less than a. Then we find the probability that the value is less than b. If we subtract these 2 probabilities, we will get the probability that z is between a and b.

Example 8: The weight of a bag of peanuts at a certain baseball game is normally distributed with an average weight of 8 ounces and a standard deviation of 0.2 ounces. What is the probability that a bag will weigh between 7.75 and 8.15 ounces?

First, we compute the z score for a weight of less than 7.75 ounces. We get $z = \frac{7.75 - 8}{0.2} = -1.25$. Let's make a sketch to see what this looks like on a bell curve:

Figure 18

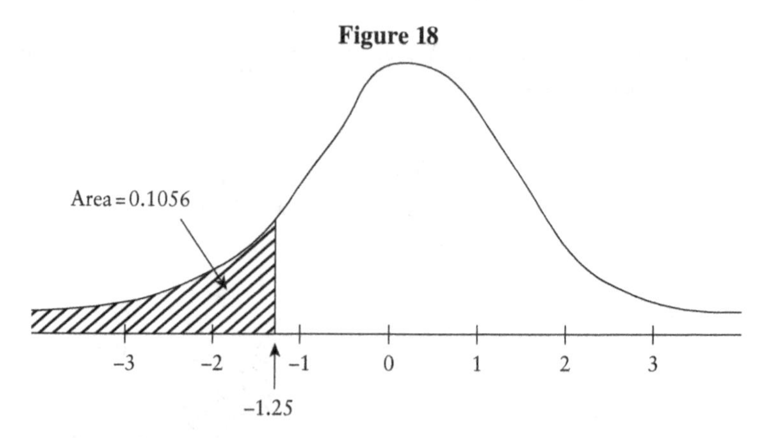

Now we go to the z table in the book. We go down the left column until we get to −1.2. Then we go across to the column under the heading 0.05 to find $P(z \leq -1.25) = 0.1056$. In other words, the probability that a bag weighs less than 7.75 ounces is 0.1056.

Now, let's compute the z score for a weight of less than 8.15 ounces. We get $z = \dfrac{8.15 - 8}{0.2} = 0.75$. Let's sketch this as well:

Figure 19

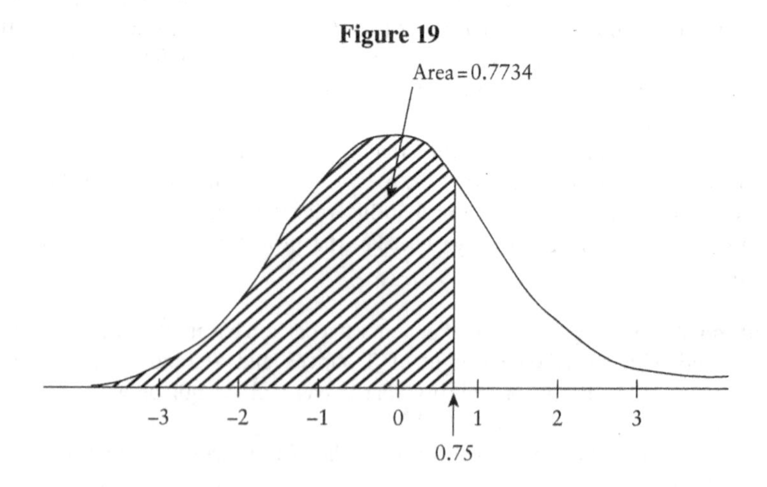

Now we go to the z table to find $P(z \leq 0.75) = 0.7734$. Now, if we find the difference between the 2 values, we get that the probability that a bag of peanuts weighs between 7.75 and 8.15 ounces is $0.7734 - 0.1056 = 0.6678$.

Figure 20

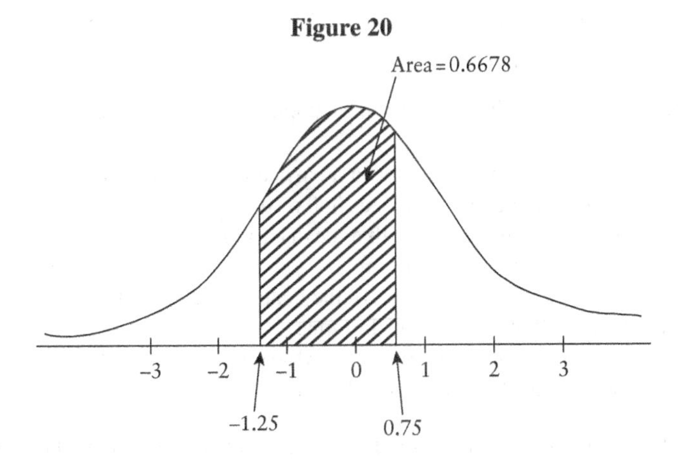

Remember that we use the statistics, μ and σ, for a population. In reality, we will rarely know these for the whole population. More often, we will take a sample of some large population and will want to do our analyses based on the sample (or samples). Then, instead of the probability distribution that we have been looking at, we will instead use a *sampling probability distribution*, which is the probability distribution of the sample means with all of the samples being of the same size n. This leads us to a crucial foundation of statistics, called the *Central Limit Theorem.* We will not prove this theorem, but let's discuss what it means.

The *Central Limit Theorem* says that if the sample size is large enough, then the distribution of sample means can be approximated with a normal distribution, *even if* the original population is not normally distributed.

Or, to be a bit more technical, The Central Limit Theorem is for a random variable x with a mean of μ and a standard deviation of σ, which are not necessarily normally distributed. All of the random samples are of the same size n, and are selected from a population, where all of the samples have the same probability of being selected. The Central Limit Theorem says:

- The distribution of sample means $\bar{x}$ will approach a normal distribution as the sample size increases.
- The mean of all of the sample means will be the population mean μ.
- The standard deviation of all of the sample means will be $\dfrac{\sigma}{\sqrt{n}}$ (not σ).

As a rule of thumb, once the sample size is greater than $n = 30$, the normal distribution will be a pretty good approximation, unless the original population is very far from normal. Also, the approximation will get better as n increases.

We will use a slightly different notation for sampling distributions. The mean is denoted $\mu_{\bar{x}}$, where $\mu_{\bar{x}} = \mu$ and the standard deviation is denoted $\sigma_{\bar{x}}$, where $\sigma_{\bar{x}} = \dfrac{\sigma}{\sqrt{n}}$. This is often called the *standard deviation of the mean.*

How will we know when to use the Central Limit Theorem?

If we are working with an individual value from a population that is normally distributed, then we will use $z = \frac{x-\mu}{\sigma}$.

If we are working with a mean from a sample, then we will use $z = \frac{\bar{x}-\mu}{\sigma/\sqrt{n}}$.

Example 9: A shot of espresso coffee at a particular coffee bar will be considered too strong if its caffeine content exceeds 350 milligrams. The coffee bar's shots of espresso are normally distributed with a mean caffeine content of 338 milligrams, and a standard deviation of 18 milligrams.

(a) Find the probability that the caffeine content of a shot of espresso selected at random will exceed 350 milligrams.

(b) If we randomly select 16 shots, what is the probability that the mean caffeine content of these shots will exceed 350?

(a) We are looking at an individual shot from a normally distributed population, so we find the z score by $z = \frac{x-\mu}{\sigma}$. Here, we get $z = \frac{350-338}{18} = 0.67$. If we look this up on the z table, we find that the probability of a shot having less than 350 milligrams of caffeine is 0.7486, and thus the probability that a shot has more than 350 milligrams of caffeine is 1 – 0.7486 = 0.2514.

(b) Now, we are looking at a randomly chosen sample of shots so we can use the Central Limit Theorem. Here, $\mu_{\bar{x}} = \mu = 338$ and $\sigma_{\bar{x}} = \frac{\sigma}{\sqrt{n}} = \frac{18}{\sqrt{16}} = 4.5$. Now if we calculate the z score, we get $z = \frac{350-338}{4.5} = 2.67$. If we look this up on the z table, we find that the probability of the mean of the sample being less than 350 milligrams of caffeine is 0.9962, and thus the probability that that the mean of the sample exceeds 350 milligrams of caffeine is 1 – 0.9962 = 0.0038.

Example 10: The mean height of a man in the United States is 69.1 inches with a standard deviation of 2.82 inches. Assume that the heights are normally distributed. If a sample of n = 50 men are chosen, what is the probability that the mean height of the sample is 72 inches (6 feet) tall?

We are looking at a randomly chosen sample of men so we can use the Central Limit Theorem. Here, $\mu_{\bar{x}} = \mu = 69.1$ and $\sigma =$ 0.3988. Now if we calculate the z score, we get $z = \frac{72-69.1}{0.3988} = 7.27$. If we look this up on the z table, we find that the probability of the mean of the men in the sample being taller than 72 is not even on the chart but is essentially 0.

What do the last 2 examples tell us? They say that the probability of an individual score deviating from the mean is higher than the probability that the mean of a sample will deviate from the mean. This should make intuitive sense. Think of the last example. You might encounter someone taller than 72 inches relatively often. After all, that is only 6 feet tall. But think about the probability that you randomly choose 50 men and that their *mean* height is 6 feet tall. That is very unlikely! In fact, if you did choose 50 men and their mean height was greater than 6 feet, you should be suspicious that you did not choose the sample randomly. We will explore this topic further when we cover hypothesis testing later in the book. And now, it's time to practice!

Practice Problems

Practice Problem 1: A quiz consists of 10 questions, each of which has 4 choices. Find the probability distribution of the number of right answers on the quiz.

Practice Problem 2: A pair of dice is rolled 6 times. Find the probability distribution of the number of sevens that could be rolled.

Practice Problem 3: Calculated the expected value for the number of correct answers in Practice Problem 1.

Practice Problem 4: Calculated the expected value for the number of sevens in Practice Problem 2.

Practice Problem 5: A game has the following rules. You place a bet of $10 and then roll a pair of dice. If you roll a 2 or a 12, you win $20. If you roll a 7, you get your $10 back. If you roll anything else, you lose your $10. What is the expected value of your winnings?

Practice Problem 6: A life insurance will pay $250,000 if the recipient dies within a 5-year period. The policy costs $500. A 22-year-old woman has a 0.9983 probability of living another 5 years. If a 22-year-old woman purchases the policy, what is its expected value?

Practice Problem 7: Suppose that 10% of the jelly beans in a jumbo bag are cherry. A random sample of 60 jelly beans are chosen. Find the mean and the standard deviation of the number of cherry jelly beans in a sample of 60 jelly beans.

Practice Problem 8: Suppose that a vaccine for a disease is 99% effective. That is, the probability that a person will not get the disease is 0.99. A random sample of 400 patients is given the vaccine. Find the mean and the standard deviation of a sample of 400 patients who are given the vaccine who can be expected to get the disease. Would it be unusual for five people to get the vaccine and then get the disease?

Practice Problem 9: Suppose that the body temperature of a human is normally distributed with a mean of 98° and a standard deviation of 0.64°. What is the probability that a human's body temperature is less than 97°?

Practice Problem 10: Suppose that the lifetime of an incandescent light bulb is normally distributed with a mean of 1100 hours and a standard deviation of 140 hours. What is the probability that an incandescent bulb will last longer than 1350 hours?

Practice Problem 11: Suppose that the weight of a beefsteak tomato is normally distributed with a mean of 14 ounces and a standard deviation of 3.2 ounces. What is the probability that a beefsteak tomato will weigh between 10 and 16 ounces?

Practice Problem 12: Suppose that a 6-ounce bowl of breakfast cereal will be considered too sweet if it contains more than 400 grams of sugar. The caffeine content of the bowls of breakfast cereal are normally distributed with a mean caffeine content of 360 grams of sugar and a standard deviation of 22 grams.

(a) Find the probability that a bowl of breakfast cereal selected at random will exceed 400 grams of sugar.

(b) If we randomly select 18 bowls, what is the probability that the mean sugar content of these bowls will exceed 400 grams of sugar?

Practice Problem 13: Suppose that the width of an adult woman's hand is normally distributed with a mean of 3.2 inches and a standard deviation of 0.6 inches.

(a) Find the probability that the mean width of an adult woman's hand will exceed 3.6 inches?

(b) If we randomly select 8 adult woman, what is the probability that the mean width of these women's hands will exceed 3.6 inches?

Practice Problem 14: Suppose that the life expectancy of a parakeet is normally distributed with a mean life expectancy of 12 years and a standard deviation of 2.2 years. If we randomly select 16 parakeets, what is the probability that the mean life expectancy of these parakeets will be less than 110 years?

Practice Problem 15: Suppose that the mean score on the bar exam for a particular state is 670 points with a standard deviation of 19 points. If we randomly select 20 people who took the bar exam for that state, what is the probability that the mean score for that group will exceed 680 points?

Solutions to Practice Problems

Solution to Practice Problem 1: *A quiz consists of 10 questions, each of which has 4 choices. Find the probability distribution of the number of right answers on the quiz.*

We can find the probability that one gets a given number of questions right using the binomial probability formula that we learned in Unit Three. Recall that the probability that a binomial event will occur r times out of a possible n is $P(r) = {}_nC_r(p^r)(q^{n-r})$, where:

n is the total number of events;

r is the number of desired events;

p is the probability that the desired event occurs;

q is the probability that the desired event does *not* occur (that is, $1 - p$).

There are 4 choices for each question, so the probability of getting a question correct is $\frac{1}{4} = 0.25$.

Here, $n = 10$, $p = 0.25$, $q = 1 - 0.25 = 0.75$, and n varies from 0 to 10.

Let's find all of the probabilities. The probabilities that one gets r questions correct are:

$$P(0) = {}_{10}C_0(0.25^0)(0.75^{10}) = 0.0563$$

$$P(1) = {}_{10}C_1(0.25^1)(0.75^9) = 0.1877$$

$$P(2) = {}_{10}C_2(0.25^2)(0.75^8) = 0.2816$$

$$P(3) = {}_{10}C_3(0.25^3)(0.75^7) = 0.2503$$

$$P(4) = {}_{10}C_4(0.25^4)(0.75^6) = 0.1460$$

$$P(5) = {}_{10}C_5(0.25^5)(0.75^5) = 0.0584$$

$$P(6) = {}_{10}C_6(0.25^6)(0.75^4) = 0.0162$$

$$P(7) = {}_{10}C_7(0.25^7)(0.75^3) = 0.0031$$

$$P(8) = {}_{10}C_8(0.25^8)(0.75^2) = 0.00039$$

$$P(9) = {}_{10}C_9(0.25^9)(0.75^1) = 0.0000286$$

$$P(10) = {}_{10}C_{10}(0.25^{10})(0.75^0) = 0.000000954.$$

Here is a table of the probabilities:

r	$P(r)$
0	0.0563
1	0.1877
2	0.2816
3	0.2503
4	0.1460
5	0.0584
6	0.0162
7	0.0031
8	0.00039
9	0.0000286
10	0.000000954

Solution to Practice Problem 2: *A pair of dice is rolled 6 times. Find the probability distribution of the number of sevens that could be rolled.*

We can find the probability that one gets a 7 on a roll using the binomial probability formula that we learned in Unit Three. Recall that the probability that a binomial event will occur r times out of a possible n is $P(r) = {}_nC_r(p^r)(q^{n-r})$, where:

n is the total number of events;

r is the number of desired events;

p is the probability that the desired event occurs;

q is the probability that the desired event does *not* occur (that is $1 - p$).

There are 6 ways to roll a 7: {(1,6), (2,5), (3,4), (4,3), (5,2), (6,1)}. There are 36 possible outcomes when one rolls a pair of dice (see Unit One), so the probability of getting a 7 is $\frac{6}{36} = \frac{1}{6}$.

Here, $n = 6$, $p = \frac{1}{6}$, $q = 1 - \frac{1}{6} = \frac{5}{6}$, and n varies from 0 to 6.

Let's find all of the probabilities. The probability that one gets r sevens is:

$$P(0) = {}_6C_0\left(\frac{1}{6}\right)^0\left(\frac{5}{6}\right)^6 = \frac{15625}{46656} \approx 0.3349$$

$$P(1) = {}_6C_1\left(\frac{1}{6}\right)^1\left(\frac{5}{6}\right)^5 = \frac{18750}{46656} \approx 0.4019$$

$$P(2) = {}_6C_2\left(\frac{1}{6}\right)^2\left(\frac{5}{6}\right)^4 = \frac{9375}{46656} \approx 0.2009$$

$$P(3) = {}_6C_3\left(\frac{1}{6}\right)^3\left(\frac{5}{6}\right)^3 = \frac{2500}{46656} \approx 0.0536$$

$$P(4) = {}_6C_4\left(\frac{1}{6}\right)^4\left(\frac{5}{6}\right)^2 = \frac{375}{46656} \approx 0.0080$$

$$P(5) = {}_6C_5\left(\frac{1}{6}\right)^5\left(\frac{5}{6}\right)^1 = \frac{30}{46656} \approx 0.000643$$

$$P(6) = {}_6C_6\left(\frac{1}{6}\right)^6\left(\frac{5}{6}\right)^0 = \frac{1}{46656} \approx 0.0000214.$$

Here is a table of the probabilities:

r	$P(r)$
0	0.3349
1	0.4019
2	0.2009
3	0.0536
4	0.0080
5	0.000643
6	0.0000214

Solution to Practice Problem 3: *Calculated the expected value for the number of correct answers in Practice Problem 1.*

Let's use the formula for expected value. We need to multiply each outcome by the probability of the outcome and sum them. We get

$$\begin{aligned} E = {} & (0)(0.0563) + (1)(0.1877) + (2)(0.2816) + (3)(0.2503) + (4)(0.1460) + \\ & (5)(0.0584) + (6)(0.0162) + (7)(0.0031) + (8)(0.00039) + (9)(0.0000286) + \\ & (10)(0.000000954) \\ = {} & 2.5001. \end{aligned}$$

Note how close this is to $(0.25)(10) = 2.5$.

Solution to Practice Problem 4: *Calculated the expected value for the number of sevens in Practice Problem 2.*

Let's use the formula for expected value. We need to multiply each outcome by the probability of the outcome and sum them. We get:

$$\begin{aligned} E = {} & (0)(0.3349) + (1)(0.4019) + (2)(0.2009) + (3)(0.0536) + (4)(0.0080) + \\ & (5)(0.000643) + (6)(0.0000214) \\ = {} & 0.9998. \end{aligned}$$

Solution to Practice Problem 5: *A game has the following rules. You place a bet of \$10 and then roll a pair of dice. If you roll a 2 or a 12, you win \$20. If you roll a 7, you get your \$10 back. If you roll anything else, you lose your \$10. What is the expected value of your winnings?*

There are 3 probabilities – either you will roll a 2 or a 12, a 7, or something else. The probability that you will roll a 2 or a 12 is $\frac{1}{36} + \frac{1}{36} = \frac{1}{18}$. The probability that you will roll a 7 is $\frac{6}{36} = \frac{1}{6}$. The probability that you will roll something other than a 2, 7, or 12 is $\frac{28}{36} = \frac{7}{9}$ (see Unit One) . If you roll a 2 or a 12, you will win \$20. You had to bet \$10, so your winnings would be \$20 – \$10 = \$10. If you roll a

7, you get your \$10 back, so your winnings would be \$10 – \$10 = \$0. If you roll anything else, you will lose your \$10. That is, your winnings will be –\$10. If we take each outcome, multiply it by its probability, and sum the products, we will get the expected value is $(10)\left(\frac{1}{18}\right)+(0)\left(\frac{1}{6}\right)+(-10)\left(\frac{7}{9}\right)\approx -7.25$. In other words, your expected value is that you will lose \$7.25

Solution to Practice Problem 6: *A life insurance will pay \$250,000 if the recipient dies within a 5-year period. The policy costs \$500. A 22-year-old woman has a 0.9983 probability of living another 5 years. If a 22-year-old woman purchases the policy, what is its expected value?*

There are 2 possible probabilities–either the woman will live for another 5 years or she will not. The probability that she will survive for another 5 years is 0.9983. The probability that she will die is thus 1 – 0.9983 = 0.0017. If she survives, she will lose \$500. That is, her winnings are –\$500. If she dies, her winnings are \$250,000. If we take each outcome, multiply it by its probability, and sum them, we will get the expected value: (0.9983)(–500) + (0.0017)(250000) = –74.15. In other words, if she lives for another 5 years, she will lose \$74.15.

Insurance is a curious "game"–if you die, you win; if you live, you lose!

Solution to Practice Problem 7: *Suppose that 10% of the jelly beans in a jumbo bag are cherry. A random sample of 60 jelly beans are chosen. Find the mean and the standard deviation of the number of cherry jelly beans in a sample of 60 jelly beans.*

We simply use the binomial formulas, where $n = 60$, $p = 0.10$, and $q = 1 - 0.10 = 0.90$. We get

$$\mu = (60)(0.10) = 6 \text{ and } \sigma = \sqrt{(60)(0.10)(0.90)} \approx 2.32.$$

Solution to Practice Problem 8: *Suppose that a vaccine for a disease is 99% effective. That is, the probability that a person will not get the disease is 0.9. A random sample of 400 patients is given the vaccine. Find the mean and the standard deviation of a sample of 400 patients who are given the vaccine who can be expected to get the disease. Would it be unusual for 5 people to get the vaccine and then get the disease?*

The probability that one will not get the disease is 0.99, so the probability that one will get the disease is 1 – 0.99 = 0.01. We use the binomial formulas, where $n = 400$, $p = 0.01$, and $q = 0.99$. We get

$$\mu = (400)(0.01) = 4 \text{ and } \sigma = \sqrt{(400)(0.01)(0.99)} \approx 1.99.$$

Five patients fall within 1 standard deviation of the mean, so it would not be unusual for 5 people to get the disease.

Solution to Practice Problem 9: *Suppose that the body temperature of a human is normally distributed with a mean of 98° and a standard deviation of 0.64°. What is the probability that a human's body temperature is less than 97°?*

First, we compute the z score. We get $z = \frac{97 - 98}{0.64} = -1.56$. Let's make a sketch to see what this looks like on a bell curve:

Figure 21

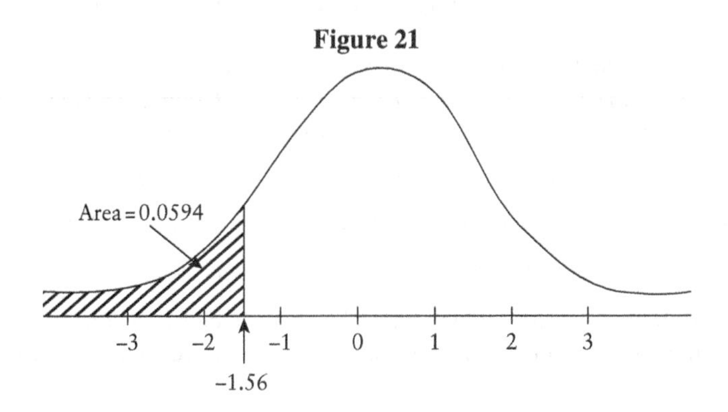

Now we go to the z table in the book and we go down the left column until we get to –1.5. Now we go across to the column under the heading 0.06, which gives us the z score for –1.56, namely $P(z \leq -1.56) = 0.0594$. In other words, the probability that a human's body temperature is less than 97° is 0.0594.

Solution to Practice Problem 10: *Suppose that the lifetime of an incandescent light bulb is normally distributed with a mean of 1100 hours and a standard deviation of 140 hours. What is the probability that an incandescent bulb will last longer than 1350 hours?*

First, we compute the z score. We get $z = \frac{1350 - 1100}{140} = 1.79$. Let's make a sketch to see what this looks like on a bell curve:

Figure 22

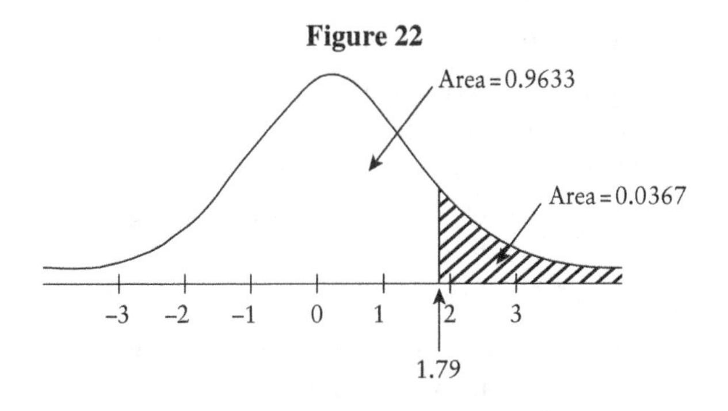

Now we go to the z table in the book and get the probability for 1.79, namely $P(z \leq 1.79) = 0.9633$, which means that the probability that a bulb will last longer than 1350 hours is $1 - 0.9633 = 0.0367$.

Solution to Practice Problem 11: *Suppose that the weight of a beefsteak tomato is normally distributed with a mean of 14 ounces and a standard deviation of 3.2 ounces. What is the probability that a beefsteak tomato will weigh between 10 and 16 ounces?*

If we want to find the probability that a tomato will weigh between 10 and 16 ounces, we find the probability that it weighs less than 10 ounces and that it weighs less than 16 ounces, and then find the difference between the 2 probabilities.

The z score for a tomato that weighs less than 10 ounces is $z = \dfrac{10-14}{3.2} = -1.25$, so the probability that a tomato will weigh less than 10 ounces is 0.1056.

The z score for a tomato that weighs less than 16 ounces is $z = \dfrac{16-14}{3.2} = 0.625$, so the probability that a tomato will weigh less than 16 ounces is 0.7324.

Therefore, the probability that a beefsteak tomato will weigh between 10 and 16 ounces is $0.7324 - 0.1056 = 0.6268$.

Solution to Practice Problem 12: *Suppose that a 6-ounce bowl of breakfast cereal will be considered too sweet if it contains more than 400 grams of sugar. The caffeine content of the bowls of breakfast cereal are normally distributed with a mean caffeine content of 360 grams of sugar and a standard deviation of 22 grams.*

(a) *Find the probability that a bowl of breakfast cereal selected at random will exceed 400 grams of sugar.*

(b) *If we randomly select 18 bowls, what is the probability that the mean sugar content of these bowls will exceed 400 grams?*

(a) The z score that a bowl of cereal has less than 400 grams of sugar is $z = \dfrac{400-360}{22} = 1.82$, so the probability that a bowl will have less than 400 grams of sugar is 0.9656, and thus that a bowl will have more than 400 grams of sugar is $1 - 0.9656 = 0.0344$.

(b) Now, we are looking at a randomly chosen sample of bowls of cereal so we can use the Central Limit Theorem. Here, $\mu_{\bar{x}} = \mu = 360$ and $\sigma_{\bar{x}} = \dfrac{\sigma}{\sqrt{n}} = \dfrac{22}{\sqrt{18}} \approx 5.185$. Now if we calculate the z score, we get $z = \dfrac{400-360}{5.185} = 7.71$. If we look this up on the z table, we find that the probability of the mean of the sample being less than 400 is not even on the table and is thus essentially 1.000 0.9962. Therefore, the probability that that the mean of the sample exceeds 400 grams of sugar is essentially 0.

Solution to Practice Problem 13: *Suppose that the width of an adult woman's hand is normally distributed with a mean of 3.2 inches and a standard deviation of 0.6 inches.*

(a) Find the probability that the mean width of an adult woman's hand will exceed 3.6 inches?

(b) If we randomly select 8 adult woman, what is the probability that the mean width of these women's hands will exceed 3.6 inches?

(a) The z score that an adult woman's hand width is greater than 3.6 inches is $z = \frac{3.6-3.2}{0.6} = 0.67$, so the probability that the width of a woman's hand will be less than 3.6 inches is 0.7486. The probability that it will be greater than 3.6 inches is $1 - 0.7486 = 0.2514$.

(b) Now, we are looking at a randomly chosen sample of adult women so we can use the Central Limit Theorem. Here, $\mu_{\bar{x}} = \mu = 4$ and $\sigma_{\bar{x}} = \frac{\sigma}{\sqrt{8}} = \frac{0.6}{\sqrt{8}} \approx 0.212$. Now if we calculate the z score, we get $z = \frac{3.6-3.2}{0.212} = 1.89$. If we look this up on the z table, we find that the probability of the mean of the sample being less than 3.6 is 0.9706. Therefore, the probability that that the mean of the sample exceeds 3.6 inches is $1 - 0.9706 = 0.0294$.

Solution to Practice Problem 14: *Suppose that the life expectancy of a parakeet is normally distributed with a mean life expectancy of 12 years and a standard deviation of 2.2 years. If we randomly select 16 parakeets, what is the probability that the mean life expectancy of these parakeets will be less than 11 years?*

We are looking at a randomly chosen sample of parakeets, so we can use the Central Limit Theorem. Here, $\mu_{\bar{x}} = \mu = 12$ and $\sigma_{\bar{x}} = \frac{\sigma}{\sqrt{n}} = \frac{2.2}{\sqrt{16}} = 0.55$. Now if we calculate the z score, we get $z = \frac{11-12}{0.55} = -1.82$. If we look this up on the z table, we find that the probability of the mean of the sample being less than 11 years is 0.0344.

Solution to Practice Problem 15: *Suppose that the mean score on the Bar exam for a particular state is 670 points with a standard deviation of 19 points. If we randomly select 20 people who took the Bar exam for that state, what is the probability that the mean score for that group will exceed 680 points?*

We are looking at a randomly chosen sample of people, so we can use the Central Limit Theorem. Here, $\mu_{\bar{x}} = \mu = 670$ and $\sigma_{\bar{x}} = \frac{\sigma}{\sqrt{n}} = \frac{19}{\sqrt{20}} = 4.25$. Now if we calculate the z score, we get $z = \frac{680-670}{4.25} = 2.35$. If we look this up on the z table, we find that the probability of the mean of the sample being less than 680 is 0.9906 and thus the probability that the mean score will exceed 680 is $1 - 0.9906 = 0.0094$.

UNIT TEN

Confidence Intervals

Now that we have learned how to find the probability of an outcome, we will work some more with estimates and samples. Remember that, the Central Limit Theorem tells us that a sample can give very good information about a population. We can use the proportion of the sample, which will generally be the best estimate of the overall proportion of the population. However, even though this might be the *best* estimate, it would be nice to know how good our estimate is. We can do this with a *confidence interval.* A *confidence interval* is an interval of values used to estimate the true valuation of a population parameter. It has an associated level of confidence, such as 90% or 95% (0.90 or 0.95), which tells us a measure of the success of the method used to find the confidence interval. There is often a probability given as α, where the confidence level is $1 - \alpha$. For example, a confidence level of 0.95 has an $\alpha = 0.05$. The confidence level is the probability that the confidence interval actually contains the population parameter if we were to repeat the estimation many times. For example, a 90% confidence interval from 0.65 to 0.85 says that we are 90% confident that the interval contains the true proportion of a population. Note that it does *not* say that there is a 90% probability that the true value lies in that interval.

How do we find a confidence interval? First, we take the critical value α and divide it in half to get $\frac{\alpha}{2}$. This is because if we are using, say, a 90% confidence interval, then there is a 10% chance that the interval does not contain the true proportion. Since the true proportion could lie above or below the interval, we want the critical value to be 5% on either side.

Figure 23

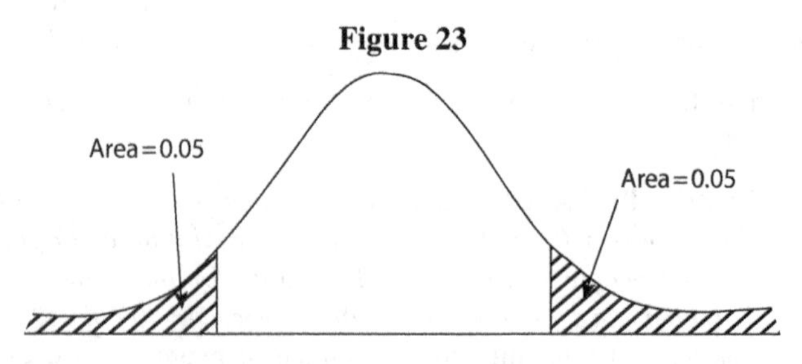

Now that we have found $\frac{\alpha}{2}$, we find the associated z score and use it to construct a *margin of error* for the confidence level. The *margin of error* for proportions is $E = z_{\alpha/2}\sqrt{\frac{\hat{p}\hat{q}}{n}}$, where $\hat{p}$ is the sample proportion that has the desired outcome,

$1-\hat{p}=\hat{q}$, and $z_{\alpha/2}$ is the z score that gives us the appropriate probability (e.g., 0.90). The confidence interval is then expressed either as $\hat{p}\pm E$ or $(\hat{p}-E, \hat{p}+E)$.

Here are 3 handy critical values:

Confidence Level	α	**Critical Value** $z_{\alpha/2}$
90%	0.10	1.645
95%	0.05	1.96
99%	0.01	2.575

We find the z score by going to the z table and working backwards. For example, to find the critical value $z_{a/2}$ for $\alpha = 0.95$, we are looking for a probability of 0.975 from the z table. We get 1.96. This will mean that the probability is 0.975 that $\hat{p}$ is below a z score of 1.96. Thus, the probability that $\hat{p}$ is above a z score of 1.96 is 1 – 0.975 = 0.025. If we also use the score of –1.96, then the probability is 0.025 that $\hat{p}$ is below –1.96.

Confused? Let's do a couple of examples.

Example 1: Suppose that 400 adults in the United States were randomly selected and asked if they have a landline in their home, and 72% said yes. Find a 90% confidence interval of the population proportion.

If we are to find a 90% confidence interval (or a critical value of $\alpha = 0.10$), we use $z_{0.05} = 1.645$ and $\hat{p} = 0.72$. Then, the margin of error is $E = 1.645\sqrt{\frac{(0.72)(0.28)}{400}} = 0.0369$. This means that our confidence interval for the proportion of adults who have a landline is 0.72 ± 0.0369. We could also write the confidence interval using another notation, (0.683, 0.757). That is, we are 90% confident that the interval from 0.683 to 0.757 contains the true proportion of adults who have a landline.

Example 2: Suppose that we surveyed 380 randomly chosen college students and found that 22% do not purchase any textbooks. Construct a 95% confidence interval for the proportion of the college student population that does not purchase any textbooks.

If we are to find a 95% confidence interval (or a critical value of $\alpha = 0.05$), we use $z_{0.025} = 1.96$ and $\hat{p} = 0.22$. Then, the margin of error is $E = 1.96\sqrt{\frac{(0.22)(0.78)}{380}} = 0.0417$. This means that our confidence interval for the proportion of college students who do not purchase any textbooks is 0.22 ± 0.0417 or (0.178, 0.262). That is, we are 95% confident that the interval from 0.178 to 0.262 contains the true proportion of college students who do not purchase any textbooks.

If we take the margin of error equation and solve it for n, we will obtain the sample size that we need to obtain a particular margin of error. We get $n = \frac{\left(z_{\alpha/2}\right)^2 \hat{p}\hat{q}}{E^2}$. Of course, it is not really necessary to memorize this formula. One could always take the margin of error equation, put in the known quantities, and solve for n. Why would we want to find n? Suppose we want to conduct a survey and we want the margin of error to be no more than 3%. How large should our sample be? By the way, if we get an answer that is not an integer, we will always round up to a larger sample size. Think about it!

Example 3: We want to conduct the survey in Example 2 and we wish to be 95% confident that our sample percentage error is no more than 3%. How many students should we survey?

Here, $\hat{p} = 0.22$, $z_{0.025} = 1.96$, and $E = 0.04$. Now we plug into the equation for sample size and we get $n = \frac{(1.96)^2 (0.22)(0.78)}{(0.04)^2} \approx 412.0116$. This means that we should survey at least 413 students.

Now that we have looked at confidence intervals for population proportions, let's look at them for a population mean. By the way, this is for a population mean where we know the standard deviation and the sample is large. In a little bit, we will look at what we do if we do not know the standard deviation or if we have a small sample (or both). Now, we will construct a confidence interval in the same way as before, except we will use $\overline{x}$ instead of $\hat{p}$. We can find E by $E = z_{\alpha/2} \cdot \frac{\sigma}{\sqrt{n}}$. That is, the confidence interval will be $\overline{x} \pm E$.

Example 4: Suppose that we are manufacturing ball bearings and we wish to know the mean radius of a ball bearing. We take a simple random sample of $n = 120$, and we determine that $\overline{x} = 0.7$ cm and $\sigma = 0.02$ cm. Find the margin of error and a 99% confidence interval for μ (the population mean).

First, let's find E. Our confidence level is 0.99 so $\alpha = 0.01$, and thus $z_{0.005} = 2.575$. We get $E = z_{0.005} \cdot \frac{\sigma}{\sqrt{n}} = 2.575 \cdot \frac{0.02}{\sqrt{120}} = 0.0047$ cm. Now we can find the confidence interval as $\mu = 0.7 \pm 0.0047$. That is, we are 99% confident that the mean radius of a ball bearing is $0.695 < \mu < 0.705$.

Let's look at Example 4 and find the 90% and 95% confidence intervals as well. First, at 90%, $E = z_{0.05} \cdot \frac{\sigma}{\sqrt{n}} = 1.645 \cdot \frac{0.02}{\sqrt{120}} = 0.003$, so the confidence interval is $0.697 < \mu < 0.703$. Now let's find the 95% confidence interval. The error $E = z_{0.025} \cdot \frac{\sigma}{\sqrt{n}} = 1.96 \cdot \frac{0.02}{\sqrt{120}} = 0.0036$, so the confidence interval is $0.696 < \mu < 0.704$.

Let's look at the three confidence intervals again:

Confidence Level	Confidence Interval
90%	$0.697 < \mu < 0.703$
95%	$0.696 < \mu < 0.704$
99%	$0.695 < \mu < 0.705$

Note that as the confidence interval becomes wider, the more confident we are. This should make sense. In other words, we are fairly confident that the mean radius of a ball bearing is between 0.697 and 0.703 cm, we are *more* confident that it is between 0.696 and 0.704 cm, and we are *very* confident that it is between 0.695 and 0.705 cm.

Just as before, we can use the margin of error to determine the sample size that we will need for estimating the population mean. If we solve the margin of error formula for n, we get $n = \frac{(z_{\alpha/2})^2\sigma^2}{E^2} = \left[\frac{z_{\alpha/2}\cdot\sigma}{E}\right]^2$.

Example 5: We wish to estimate the mean weight of college students so that we can come up with a capacity limit for the dormitory elevators. We somehow know that the standard deviation of a college student's weight is 6 pounds, and we wish to be 95% confident with a margin of error of no more than 4%. How many students should we survey?

Here, we have $\sigma = 6$, $E = 0.04$, and $z_{0.025} = 1.96$. We get $n = \left[\frac{1.96\cdot 6}{0.04}\right]^2 = 86,436$.

Note that in the above example, we come up with a necessary survey size of 86,436 students. This is a very large number! It is highly impractical (and expensive!) to survey that many students, so we could either increase the margin of error or have a lower level of confidence. More realistically, when we are sampling, we will have small sample sizes and we will not know σ. Then, we will need to do something different.

There are 2 crucial assumptions to the Central Limit Theorem, which justifies our use of the z statistic–it depends on a large sample size, and we know the standard deviation of the population. But, in most situations, the sample size will be small and we will not know the standard deviation.

Recall that when we do not know the standard deviation, we estimate the standard deviation of the sample, s. If we take the formula for the z statistic of a sample and replace σ with s, we get a new random variable, which we call t statistic. Namely, $t = \frac{\bar{x} - \mu}{s/\sqrt{n}}$.

There is a t distribution associated with the t statistic just as there was with the z distribution. The t distribution looks like the z distribution, except that it is "flatter" and has fatter tails. As the sample size gets larger, the t distribution looks more like the z distribution:

Figure 24

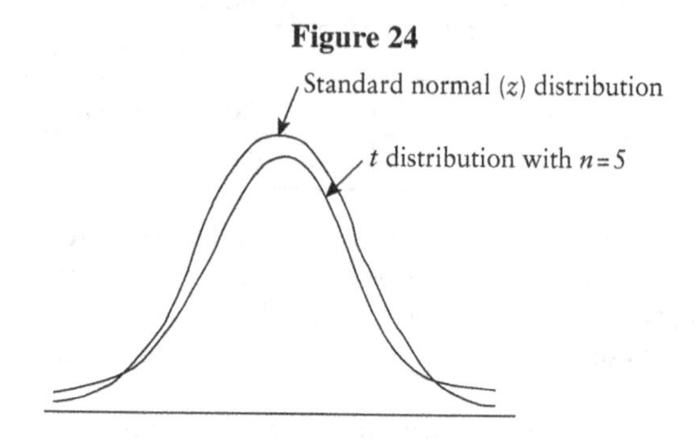

When we compute confidence intervals with small samples, we compensate for the unreliability of small samples by using this t distribution. Otherwise, the procedure is the same as before. We can find values for the t distribution in Table 2 in the back of the book. To find a critical value of $t_{\alpha/2}$ we locate the number of *Degrees of Freedom* in the left column, then read across the appropriate row until reaching the number corresponding to the desired area at the top. The number of *degrees of freedom* of a sample of size n is $d.f. = n-1$. Why is this true? Imagine that we have a sample of n scores. If we want to have a particular mean, then we can give $n-1$ of the scores any value that we want, but the *nth* value is forced to be a particular number. For example, let's say that we have 8 quiz scores, and we know that the mean of the 8 scores is 75. Then the total of the 8 scores must be $8 \cdot 75 = 600$. We could assign any score that we wanted to 7 of the quizzes, but in order for the mean to be 75, the eighth score must be 600 minus the total of the other 7 scores. Thus, 7 of the scores are *free.*

The margin of error, E, for the estimate of μ when σ is unknown, is $E = t_{\alpha/2} \cdot \frac{s}{\sqrt{n}}$, where $t_{\alpha/2}$ has $n-1$ degrees of freedom.

The confidence interval for the estimate of μ when σ is unknown is $\bar{x} - E < \mu < \bar{x} + E$.

Example 6: As in Example 5, we wish to estimate the mean weight of college students so that we can come up with a capacity limit for the dormitory elevators. We sample 30 of them and we get $\bar{x} = 142$ pounds with $s = 11$ pounds. We wish to be 95% confident in our estimate of the mean weight. Find the margin of error and construct the confidence interval for μ.

To find E, we will need the appropriate t score. We go to the table and look at $30 - 1 = 29$ degrees of freedom. Then we read across to $\alpha = 0.025$. We get $t_{0.025} = 2.045$. Thus the margin of error is $E = 2.045 \cdot \frac{11}{\sqrt{30}} = 4.107$.

Now we can find the confidence interval, using $\overline{x}-E<\mu<\overline{x}+E$. We get 142−4.107< μ < 142 + 4.107, or (137.89, 146.11). Thus, we are 95% confident that our estimate of the mean weight of the students is between 137.89 pounds and 146.11 pounds.

How do we know whether to use the z or t distribution?

If we know σ:

- If the population is normally distributed, we use the z distribution.
- If the population is not normally distributed but $n > 30$, we can still use the z distribution.
- If the population is not normally distributed and $n \leq 30$, we need to use methods that are not covered in this book.

If we do not know σ:

- If the population is normally distributed, we use the t distribution.
- If the population is not normally distributed but $n > 30$, we can still use the t distribution.
- If the population is not normally distributed and $n \leq 30$, we need to use methods that are not covered in this book.

In addition to estimating the mean of a sample, we will often want to estimate the variance. We can do this using a *chi-square distribution*. The chi-square distribution requires, again, that we have a simple random sample and that the population be normally distributed. We find the Chi-Square Statistic with the formula: $\chi^2 = \frac{(n-1)s^2}{\sigma^2}$, where n is the sample size, s^2 is the sample variance, and σ^2 is the population variance. The distribution looks like this:

Figure 25

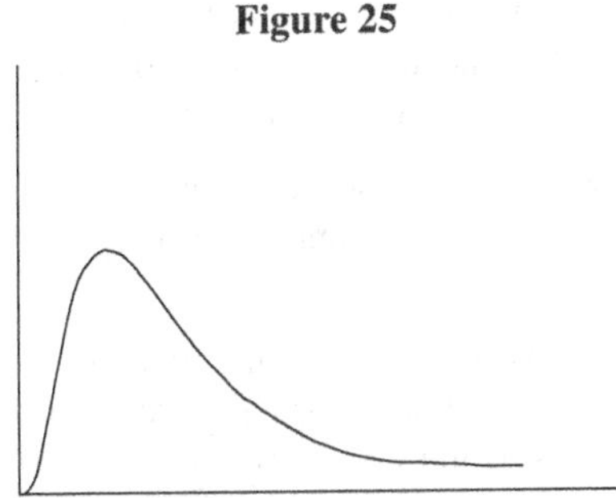

Note that the distribution is not symmetric and that the values of χ^2 can be zero or positive, but not negative. The χ^2 distribution is different for a different number of degrees of freedom (as before, $d.f. = n - 1$). The chi-square distribution (Table 3 in the back of the book) gives the area to the right of the critical value. Because the distribution is not symmetric, we will need to calculate separately the upper and lower limits of the confidence interval. We then get the confidence interval for the population variance by $\frac{(n-1)s^2}{\chi_R^2} < \sigma^2 < \frac{(n-1)s^2}{\chi_L^2}$.

Example 7: We survey the weights of 20 college students and find that the mean weight of the sample is $\overline{x} = 142$ pounds with a sample standard deviation of $s = 11$ pounds. Assume that the population has a normal distribution and that we have a simple random sample. Construct a 90% confidence interval ($\alpha = 0.10$) of the population variance.

First, we find the degrees of freedom. We get *d.f.* = 20 – 1 = 19. Now, to find the right-hand limit, we go to the chi-square table, read down the left-most column until we get to 19, and read across to the column headed 0.05. We get 30.14 Now, to get the left-hand limit, we subtract the area from 1 to get 1 – 0.05 = 0.95. Now, in the row where degrees of freedom is 19, we go to the column headed 0.95 and we get 10.12. Now we can plug our information into the formula. We get $\frac{(20-1)(11)^2}{30.14} < \sigma^2 < \frac{(20-1)(11)^2}{10.12}$, or $76.277 < \sigma^2 < 227.174$.

If we take the square roots of each side of the confidence interval, we get $8.73 < \sigma < 15.07$. Thus, we are 90% confident that the true value of the standard deviation of the weights is between 8.73 and 15.07 pounds.

Are you ready to practice?

Practice Problems

Practice Problem 1: Suppose that 300 adults in the United States were randomly selected and asked if they support candidate A for President. The results show that 54% said yes. Find a 95% confidence interval of the population proportion.

Practice Problem 2: Suppose we randomly select and ask 80 college students to name their favorite sport. We find that 62% answer football. Find a 90% confidence interval of the proportion of the college student population.

Practice Problem 3: We want to conduct the survey in Practice Problem 1, and we wish to be 95% confident that our sample percentage error is no more than 3%. How many adults should we survey?

Practice Problem 4: We want to conduct the survey in Practice Problem 2, and we wish to be 90% confident that our sample percentage error is no more than 2%. How many college students should we survey?

Practice Problem 5: Suppose that we are manufacturing chocolate bars, and we wish to know the mean weight of a chocolate bar. We take a simple random sample of $n = 60$, and we determine that $\overline{x} = 8.05$ ounces and $\sigma = 0.11$ ounces. Find the margin of error and a 99% confidence interval for μ (the population mean).

Practice Problem 6: We wish to estimate the mean weight of checked luggage for the airplanes in our fleet. We know that the standard deviation of an item of luggage is 4.5 pounds, and we wish to be 95% confident, with a margin of error of no more than 4%. How many items of luggage should we survey?

Practice Problem 7: As in Practice Problem 6, we wish to estimate the mean weight of checked luggage for our airplane fleet. We randomly sample 40 of them and we get $\overline{x} = 42$ pounds with $s = 4.7$ pounds. We wish to be 95% confident in our estimate of the mean weight. Find the margin of error and construct the confidence interval for μ.

Practice Problem 8: We wish to estimate the mean number of calories that the typical adult American male consumes in a day. We randomly sample 60 of them and we get $\overline{x} = 2655$ calories with $s = 244$ calories. We wish to be 99% confident in our estimate of the mean number of calories. Find the margin of error and construct the confidence interval for μ.

Practice Problem 9: We survey the calorie consumption of 100 randomly chosen adults and find that the mean consumption of the sample is $\overline{x} = 2655$ calories with a sample standard deviation of $s = 244$ calories. Assume that the population has a normal distribution and that we have a simple random sample. Construct a 95% confidence interval ($\alpha = 0.05$) of the population variance.

Practice Problem 10: We survey 30 randomly chosen college graduates and we find that the mean level of college debt is \$18,500 with a sample standard deviation of s = \$1050. Assume that the population of college graduate debt has a normal distribution and that we have a simple random sample. Construct a 99% confidence interval ($\alpha = 0.01$) of the population variance.

Solutions to Practice Problems

Solution to Practice Problem 1: *Suppose that 300 adults in the United States were randomly selected and asked if they support candidate A for President. The results show that 54% said yes. Find a 95% confidence interval of the population proportion.*

If we are to find a 95% confidence interval (or a critical value of $\alpha = 0.05$), we use $z_{0.025} = 1.96$ and $\hat{p} = 0.54$. Then the margin of error is $E = 1.96\sqrt{\frac{(0.54)(0.46)}{300}} = 0.0564$. This means that our confidence interval for the proportion of adults who support candidate A is 0.54 ± 0.0564, or (0.484, 0.596). That is, we are 95% confident that the interval from 0.484 to 0.596 contains the true proportion of adults who favor candidate A for President.

Solution to Practice Problem 2: *Suppose we randomly select and ask 80 college students to name their favorite sport. We find that 62% answer football. Find a 90% confidence interval of the proportion of the college student population.*

If we are to find a 90% confidence interval (or a critical value of $\alpha = 0.10$), we use $z_{0.05} = 1.645$ and $\hat{p} = 0.62$. Then the margin of error is $E = 1.645\sqrt{\frac{(0.62)(0.38)}{80}} = 0.0893$. This means that our confidence interval for the proportion of college students whose favorite sport is football is 0.62 ± 0.0893, or (0.531, 0.709). That is, we are 90% confident that the interval from 0.531 to 0.709 contains the true proportion of college students whose favorite sport is football.

Solution to Practice Problem 3: *We want to conduct the survey in Practice Problem 1, and we wish to be 95% confident that our sample percentage error is no more than 3%. How many adults should we survey?*

Here, $\hat{p} = 0.54$, $z_{0.025} = 1.96$, and $E = 0.03$. Now we plug into the equation for sample size and we get $n = \frac{(1.96)^2(0.54)(0.46)}{(0.03)^2} \approx 1060.2816$. This means that we should survey at least 1061 students.

Solution to Practice Problem 4: *We want to conduct the survey in Practice Problem 2, and we wish to be 90% confident that our sample percentage error is no more than 2%. How many college students should we survey?*

Here, $\hat{p} = 0.62$, $z_{0.05} = 1.645$, and $E = 0.02$. Now we plug into the equation for sample size and we get $n = \frac{(1.645)^2(0.62)(0.38)}{(0.02)^2} \approx 1593.8487$. This means that we should survey at least 1594 students.

Solution to Practice Problem 5: *Suppose that we are manufacturing chocolate bars, and we wish to know the mean weight of a chocolate bar. We take a simple random sample of n = 60, and we determine that $\bar{x} = 8.05$ ounces and $\sigma = 0.11$ ounces. Find the margin of error and a 99% confidence interval for μ (the population mean).*

First, let's find E. Our confidence level is 0.99 so $\alpha = 0.01$, and thus $z_{0.005} = 2.575$. We get $E = z_{0.005} \cdot \frac{\sigma}{\sqrt{n}} = 2.575 \cdot \frac{0.11}{\sqrt{60}} = 0.03657$ ounces. Now we can find the confidence interval: $\mu = 8.05 \pm 0.03657$. That is, $8.01343 < \mu < 8.08657$.

Solution to Practice Problem 6: *We wish to estimate the mean weight of checked luggage for the airplanes in our fleet. We know that the standard deviation of an item of luggage is 4.5 pounds, and we wish to be 95% confident, with a margin of error of no more than 4%. How many items of luggage should we survey?*

Here, we have σ = 4.5, E = 0.04, and $z_{0.025}$ = 1.96. We get $n = \left[\frac{(1.96)(4.5)}{0.04}\right]^2$ = 48,620.25. Therefore, we will need to survey 48,621 items of luggage.

Solution to Practice Problem 7: *As in Practice Problem 6, we wish to estimate the mean weight of checked luggage for our airplane fleet. We randomly sample 40 of them and we get $\overline{x} = 42$ pounds with $s = 4.7$ pounds. We wish to be 95% confident in our estimate of the mean weight. Find the margin of error and construct the confidence interval for μ.*

Note that we do not know σ, so we will use a t distribution, *not* a z distribution in this problem. To find E, we will need the appropriate t score. We go to the table and look at 40 – 1 = 39 degrees of freedom. If there is not an appropriate row for degrees of freedom, we should round down and use the closest degrees of freedom that are *lower* that have a larger t value than our actual degrees of freedom. Here, we have rows for 38 and 40, so we will use the row for 35 *d.f.* We read across to α = 0.025. We get $t_{0.025} = 2.030$. Thus the margin of error is $E = 2.030 \cdot \frac{4.8}{\sqrt{40}} = 1.541$.

Now we can find the confidence interval, using $\overline{x} - E < \mu < \overline{x} + E$. We get 42 – 1.541 < μ < 42 + 1.541, or (40.459, 43.541). Thus, we are 95% confident that our estimate of the mean weight of an item of luggage is between 40.459 pounds and 43.541 pounds.

Solution to Practice Problem 8: *We wish to estimate the mean number of calories that the typical adult American male consumes in a day. We randomly sample 60 of them and we get $\overline{x} = 2655$ calories with s = 244 calories. We wish to be 99% confident in our estimate of the mean number of calories. Find the margin of error and construct the confidence interval for μ.*

Note that we do not know σ, so we will use a t distribution, *not* a z distribution in this problem. To find E, we will need the appropriate t score. We go to the table and look at 60 – 1 = 59 degrees of freedom. If there is not an appropriate row for degrees of freedom, we should use the closest degrees of freedom that are *lower* that have a larger t value than our actual degrees of freedom. Here, we have rows for 50 and 60, so we will use the row for 50 *d.f.* We read across to $\alpha = 0.01$. We get $t_{0.01} = 2.403$. Thus, the margin of error is $E = 2.403 \cdot \frac{244}{\sqrt{60}} = 75.695$.

Now we can find the confidence interval, using $\overline{x} - E < \mu < \overline{x} + E$. We get 2655 – 75.695 < μ < 2655 + 75.695, or (2579.305, 2730.695). Thus, we are 99% confident that our estimate of the mean number of calories is between 2579.305 and 2730.695 calories.

Solution to Practice Problem 9: *We survey the calorie consumption of 100 randomly chosen adults and find that the mean consumption of the sample is $\overline{x} = 2655$ calories with a sample standard deviation of s = 244 calories. Assume that the population has a normal distribution and that we have a simple* *random sample. Construct a 95% confidence interval ($\alpha = 0.05$) of the population variance.*

First, we find the degrees of freedom. We get *d.f.* = 100–1 = 99. If there is not an appropriate row for degrees of freedom, we should use the closest degrees of freedom that are *lower* than our actual degrees of freedom. Here, we have rows for 90 and 100, so we will use the row for 90 *d.f.* Now, to find the right hand limit, we go to the chi-square table, read down the leftmost column until we get to 90, and read across to the column headed 0.05. We get 113.15. Now, to get the left-hand limit, we subtract the area from 1 to get 1 – 0.05 = 0.95. Now, in the row where the number of degrees of freedom is 29, we go to the column headed 0.95 and we get 69.13. Now we can plug our information into the formula. We get $\frac{(100-1)(244)^2}{113.15} < \sigma^2 < \frac{(100-1)(244)^2}{69.13}$, or $52090.711 < \sigma^2 < 85260.582$. If we take the square roots, we can estimate the standard deviation with $228.234 < \sigma < 291.994$.

Solution to Practice Problem 10: *We survey 30 randomly chosen college graduates and we find that the mean level of college debt is \$18,500 with a sample standard deviation of s = \$1050. Assume that the population of college graduate debt has a normal distribution and that we have a simple random sample. Construct a 99% confidence interval ($\alpha = 0.01$) of the population variance.*

First, we find the degrees of freedom. We get *d.f.* = 30 – 1 = 29. Now, to find the right-hand limit, we go to the chi-square table, read down the left-most column until we get to 29, and read across to the column headed 0.01. We get 49.59. Now, to get the left-hand limit, we subtract the area from 1 to get 1 – 0.01 = 0.99. Now, in the row where the number of degrees of freedom is 29, we go to the column headed 0.99 and we get 14.26. Now we can plug our information into the formula. We get $\frac{(30-1)(1050)^2}{49.59} < \sigma^2 < \frac{(30-1)(1050)^2}{14.26}$, or $644736.842 < \sigma^2 < 2242110.799$. If we take the square roots, we can estimate the standard deviation with $802.955 < \sigma < 13476.312$.

UNIT ELEVEN

Hypothesis Testing

Now we will look at an area of Statistics that you will encounter often as a basic procedure for determining the likelihood that an observation or set of observations could have occurred by chance. It is widely used, and occasionally abused, in fields like medicine, Psychology, Sociology, business, and a variety of others. A *hypothesis* is a claim or a statement that we will make about some aspect of a population. A *hypothesis test* is a procedure for testing that hypothesis, which tries to assess the probability that the observations could have easily occurred slowly by chance. There are several components to a hypothesis test.

- The *null hypothesis* is a statement that the value of a population parameter (usually a proportion, mean or standard deviation) is equal to some value. We will test the null hypotheses by assuming it is true and then concluding that we either need to reject that assumption or that we fail to reject it. The null hypothesis is usually denoted by H_0. The null hypothesis is usually that the observations occurred by chance.
- The *alternative (or alternate) hypothesis* is the statement that the population parameter is different from that of the null hypothesis: either greater than, less than, or not equal to H_0. The alternative hypothesis states the observations are the result of some effect (plus, of course, some element of chance), and not merely by chance. The alternative hypothesis is usually denoted by H_1 (or sometimes H_a).

If we are using a hypothesis test to support a claim, this claim will be the alternative hypothesis. Then we test to see whether the claim could have occurred by chance (H_0). If we can reject H_0, then we can argue that H_1 is valid. Note that we can not necessarily prove that H_1 is true, but rather that the observed effect did not occur by chance.

Suppose we wanted to test the claim that the mean amount of caffeine in a cup of our green tea is at least 50 milligrams. That is, $\mu \geq 50$. This claim would be false if $\mu < 50$, so $H_1 < 50$. But, we test $H_0 = 50$. Why? First, we are just trying to determine if there is sufficient evidence to reject the null hypothesis. If so, then we can determine if the mean amount of caffeine is at least 50 milligrams.

- The *test statistic* will be used to determine whether to reject the null hypothesis. The test statistic will either be the z score (for proportion or mean), t score (for mean), or χ^2 score (for standard deviation).
- The *critical value* is the set of all values of the test statistic that result in rejecting the null hypothesis.
- The *significance level* (α) is the probability that the test statistic is in the critical value when the null hypothesis turns out to be true.

- The *p-value* is the probability of getting a test statistic that is less extreme than the one for the sample data, presuming the null hypothesis is true.

Hypothesis testing is usually done by 1 of 2 methods:

- *Traditional Method:* We find the test statistic and reject H_0 if the test statistic falls within the critical value. Otherwise, we fail to reject H_0.
- *p-Value Method:* We find the p-value and reject H_0 if the p-value $\leq \alpha$. Otherwise, we fail to reject H_0.

We could also do a hypothesis test by finding a confidence interval and seeing if the parameter has a value that is outside of the confidence interval.

Here are more important points to keep in mind. When we fail to reject the null hypothesis, we are not proving that it is true; we are simply saying that we do not have sufficient evidence to reject the claim. In a courtroom, saying that the defendant is not guilty does not mean that he or she is innocent, but merely that the prosecution has not proved their case sufficiently.

Finally, there are 2 types of errors that we are trying to avoid.

A *Type I error* means that we rejected the null hypothesis when it was actually true.

A *Type II error* means that we failed to reject the null hypothesis when it was actually false.

The probability of a Type I error is usually symbolized by α, and the probability of a Type II error is usually symbolized by β.

Let's do some examples.

Example 1: A certain high school faculty has 50 male and 50 female teachers. Each year, the principal randomly picks 20 teachers to represent the school at graduation. Of the 20 teachers that year, only 2 were female and 18 were male. Is the principal guilty of sexism?

What we want to know is the likelihood that out of 20 teachers, the principal only selected 2 females by chance. The number of women is a binomial random variable with $n = 20$ trials and $p = 0.5$.

Let's test this hypothesis.

H_0 is the null hypothesis that the teachers are randomly chosen.

H_1 is the alternative hypothesis that there is some bias in the selection of the teachers.

The test statistic is the binomial random variable with $p = 0.5$ and $n = 20$.

The p-value is ${}_{20}C_2\,(0.5)^2(0.5)^{18} = 0.00018$.

If we were testing this at the $\alpha = 0.05$ level of significance, then we will reject the null hypothesis if $p \leq \alpha$. Here, p is much less than 0.05, so we can reject the null hypothesis and say that it is likely that the selection of teachers is not random.

Note that it is *possible* that the selection did occur by chance, but that it is very unlikely.

Example 2: Suppose that a survey of 400 randomly selected college students asked whether they had ever cheated on an exam and 73% admitted that they had.

We wish to test the claim at the $\alpha = 0.05$ level of significance that more than $\frac{2}{3}$ of college students have cheated on an exam.

Let's first test this with the traditional method and then with the p-value method.

Our null hypothesis is H_0: $p = \frac{2}{3}$ (we will use 0.667 for simplicity).

Our alternative hypothesis is H_1: $p > \frac{2}{3}$.

The test statistic is found using n = 400 and $\hat{p} = 0.73$. We get $z = \frac{0.73 - 0.667}{\sqrt{\frac{(0.5)(0.5)}{400}}} = 2.52$. The critical value will be z = 1.645. Our test statistic falls within the rejection region (it is greater than 1.645) so we can reject the null hypothesis. That is, if the fraction of college students who have cheated on an exam is less than or equal to two-thirds, it is very unlikely that we obtained 0.73 by chance.

If we use the p-value method, we find the region to the right of the test statistic. We get 0.9941. We need to get a p-value of $1 - \alpha = 1 - 0.05 = 0.95$, in order to reject the null hypothesis.

Example 3: Suppose that we randomly sample 150 dogs and find that their mean body temperature is 102°, with a known population standard deviation of 0.75°. Test at the 0.05 level of significance that the mean body temperature of a dog is actually 101°.

Let's test this with the traditional method.

Our null hypothesis is H_0: $\mu = 101$.

Our alternative hypothesis is H_1: $\mu > 101$.

The test statistic is found using n = 150 and $\bar{x} = 102$, with σ = 0.75°. We get $z = \frac{102 - 101}{0.75/\sqrt{150}} = 16.33$. The critical value will be z = 1.645. Our test statistic falls within the rejection region (it is greater than 1.645), so we can reject the null hypothesis. That is, if the mean body temperature of a dog is 101°, it is very unlikely that we could have tested 150 of them and obtained a mean temperature of 102° by chance.

If we had used the p-value method, we find the region to the right of the test statistic. We get 0.9999. We need to get a p-value of $1 - \alpha = 1 - 0.05 = 0.95$, so we can reject the null hypothesis.

Example 4: Suppose that we randomly sample 15 dogs and find that their mean body temperature is 102° with a sample standard deviation of 0.75°. Test at the 0.05 level of significance that the mean body temperature of a dog is actually 101°.

Let's test this with the traditional method.

Our null hypothesis is H_0: $\mu = 101$.

Our alternative hypothesis is H_1: $\mu > 101$.

Here, we have a simple random sample with an unknown population standard deviation. So, instead of a z score, we will find a t score. The test statistic is found using n = 15 and $\bar{x} = 102$, with s = 0.75° and *d.f.* = 15 − 1 = 14. We get $t = \frac{102-101}{0.75/\sqrt{15}} = 5.16$. The significance level is α = 0.05, and the critical value will be 1.761 (from the t table). Our test statistic falls within the rejection region (it is greater than 1.761), so we can reject the null hypothesis. That is, if the mean body temperature of a dog is 101°, it is very unlikely that we could have tested 15 of them and obtained 102° by chance.

By the way, finding p-values is more complicated with the t distribution because it requires either software or a calculator. Thus, we will use the traditional method for tests that use a t statistic.

Example 5: The scores on the math placement exam of a university are normally distributed with a mean of 74 and a standard deviation of 9. A professor decides to test this by taking a random sample of 15 scores and finds a standard deviation of 6 points. Now the professor is unsure whether the standard deviation is really 9. She decides to test the claim that $\sigma = 9$, and, because she is fussy, at the 0.01 level of significance.

Our null hypothesis is H_0: $\sigma = 9$.

Our alternative hypothesis is H_1: $\sigma \neq 9$.

The significance level is $\alpha = 0.01$. We will use the χ^2 distribution to test this claim because it is about a standard deviation.

The test statistic is $\chi^2 = \frac{(15-1)(6)^2}{9^2} = 6.222$.

The critical values are $\chi^2 = 4.075$ and $\chi^2 = 31.319$. Because the test statistic is *not* in the critical value, we *cannot* reject the null hypothesis. That is, we might have obtained a standard deviation of 9 by chance. We should now do the experiment again, with a larger sample, to see if the standard deviation differs from 9.

Are you ready to practice?

Practice Problems

Practice Problem 1: A bag of trail mix is created by a machine that chooses either almonds or raisins from a pair of very large bins, each of which containing only raisins or almonds. Each bag usually contains 40% almonds. We open a bag and find that, out of the 100 nuts in the bag, only 34 were almonds. Can we conclude that a bag of mixed nuts usually has less than 40% at a level of significance of $\alpha = 0.05$? How about at $\alpha = 0.01$?

Practice Problem 2: The student body at a university is 50% male and 50% female. In a math class of 25 students, only 10 of them are female. Test at the $\alpha = 0.05$ level of significance that this could have occurred randomly?

Practice Problem 3: Suppose that a survey of 500 randomly selected registered voters are asked whether they voted in the last presidential election and that 49%

said that they had. We wish to test the claim at the $\alpha = 0.05$ level of significance that more than 50% of registered voters voted in the last presidential election.

Practice Problem 4: Suppose that a survey of 400 randomly selected customers at a restaurant asked them whether they were happy with their last meal and 81% said that they were. We wish to test the claim at the $\alpha = 0.05$ level of significance that more than three quarters of the customers at the restaurant are happy with their meals.

Practice Problem 5: Suppose that we randomly sample 180 piston diameters and find that their mean diameter is 3 centimeters, with a known population standard deviation of 0.15 centimeter. Test at the 0.05 level of significance that the mean diameter of a piston is actually 3.25 centimeters.

Practice Problem 6: Suppose that we randomly sample 100 manufactured DVDs and find that the mean error rate in storage is 0.1%, with a known population standard deviation of 0.02%. Test that the mean error rate is actually 0.11% at the 0.01 level of significance.

Practice Problem 7: Suppose that we randomly sample 20 cups of espresso and find that their mean temperature is 155°, with a sample standard deviation of 8°. Test that the mean temperature of a cup of espresso is actually 165° at the 0.05 level of significance.

Practice Problem 8: Suppose that we randomly sample the resting pulse rate of 12 female college soccer players and obtain a mean pulse rate of 71 beats per minute, with a standard deviation of 4 beats per minute. Test that the mean pulse rate is actually 75 beats per minute at the 0.01 level of significance.

Practice Problem 9: A freshman at our university usually gains a mean of 7.2 pounds with a standard deviation of 2.1 pounds. A random sample of 21 students obtains a standard deviation of 3.2 pounds. Test the claim that $\sigma = 2.1$ at the 0.05 level of significance.

Practice Problem 10: Suppose that the mean stopping time of a particular model of car moving at 40 *mph* is listed as 1.8 seconds with a standard deviation of 0.25 seconds. A random sample of 12 cars obtains a standard deviation of 0.35 seconds. Test the claim that $\sigma = 0.25$ seconds at the 0.05 level of significance.

Solutions to Practice Problems

Solution to Practice Problem 1: *A bag of trail mix is created by a machine that chooses either almonds or raisins from a pair of very large bins, each of which containing only raisins or almonds. Each bag usually contains 40% almonds. We open a bag and find that, out of the 100 nuts in the bag, only 34 were almonds. Can we conclude that a bag of mixed nuts usually has less than 40% almonds at a level of significance of $\alpha = 0.05$. How about at $\alpha = 0.01$?*

What we want to know is the likelihood that out of 100 raisins and almonds, only 34 are almonds. The number of almonds is a binomial random variable, with $n = 100$ trials and $p = 0.4$.

Let's test the hypothesis.

H_0 is the hypothesis that a bag contains 40% almonds.

H_1 is the hypothesis that bag contains less than 40%.

The test statistic is the binomial random variable, with $p = 0.4$ and $n = 100$.

The p-value is ${}_{100}C_{34}\,(0.4)^{34}(0.6)^{66} = 0.03908$.

If we were testing this at the $\alpha = 0.05$ level of significance, then we would reject the null hypothesis because $p \leq \alpha$. If we were testing this at the $\alpha = 0.01$ level of significance, then we could *not* reject the null hypothesis.

Solution to Practice Problem 2: *The student body at a university is 50% male and 50% female. In a math class of 25 students, only 10 of them are female. Test at the $\alpha = 0.05$ level of significance that this could have occurred randomly?*

The number of females is a binomial random variable, with n = *25* trials and $p = 0.5$.

Let's test the hypothesis.

H_0 is that a class contains 50% females.

H_1 is that class contains less than 50%.

The test statistic is a binomial random variable with $p = 0.5$ and $n = 25$.

The p-value is ${}_{25}C_{10}\,(0.5)^{10}(0.5)^{15} = 0.0974$.

At the $\alpha = 0.05$ level of significance, we can *not* reject the null hypothesis because $p \geq \alpha$. If we were testing this at the $\alpha = 0.10$ level of significance, then we could have rejected the null hypothesis.

Solution to Practice Problem 3: *Suppose that a survey of 500 randomly selected registered voters are asked whether they voted in the last presidential election and that 49% said that they had. We wish to test the claim at the $\alpha = 0.05$ level of significance that more than 50% of registered voters voted in the last presidential election.*

Let's first test this with the traditional method and then with the p-value method.

Our null hypothesis is H_0: $p = 0.50$.

Our alternative hypothesis is H_1: $p < 0.50$.

The test statistic is found using $n = 500$ and $\hat{p} = 0.49$. We get $z = \dfrac{0.49 - 0.50}{\sqrt{\dfrac{(0.5)(0.5)}{500}}} = -0.4472$. The critical value will be $z = -1.645$. Our test statistic does not fall within the rejection region (it is greater than -1.645), so we cannot reject the null hypothesis.

If we use the p-value method, we find the region to the left of the test statistic. We get 0.3264. We need to get a p-value of $\alpha = 0.05$, so we cannot reject the null hypothesis.

Solution to Practice Problem 4: *Suppose that a survey of 400 randomly selected customers at a restaurant asked them whether they were happy with their last meal* *and 81% said that they were. We wish to test the claim at the $\alpha = 0.05$ level of significance that more than three quarters of the customers at the restaurant are happy with their meals.*

Let's first test this with the traditional method and then with the p-value method.

Our null hypothesis is H_0: $p = 0.75$

Our alternative hypothesis is H_1: $p > 0.75$

The test statistic is found using $n = 100$ and $\hat{p} = 0.81$. We get $z = \frac{0.81 - 0.75}{\sqrt{\frac{(0.75)(0.25)}{400}}} = 2.77$. The critical value will be $z = 1.645$. Our test statistic falls within the rejection region (it is greater than 1.645) so we can reject the null hypothesis.

If we use the p-value method, we find the region to the right of the test statistic. We get 0.9972. We need to get a p-value of $1 - \alpha = 1 - 0.05 = 0.95$, so we can reject the null hypothesis.

Solution to Practice Problem 5: *Suppose that we randomly sample 150 piston diameters and find that their mean diameter is 3 centimeters, with a known population standard deviation of 0.15 centimeter. Test that the mean diameter of a piston is actually 3.25 centimeters at the 0.01 level of significance.*

Let's test this with the traditional method.

Our null hypothesis is H_0: $\mu = 3.25$.

Our alternative hypothesis is H_1: $\mu < 3.25$.

The test statistic is found using $n = 150$ and $\bar{x} = 3$ with $\sigma = 0.15$. We get $z = \frac{3 - 3.25}{0.15/\sqrt{150}} = -20.41$. The critical value will be $z = -1.645$. Our test statistic falls within the rejection region (it is less than -1.645) so we can reject the null hypothesis. That is, if the mean diameter of a piston is 3.25 centimeters, then it is very unlikely that we could have tested 150 of them and obtained a mean diameter of 3 centimeters by chance.

If we had used the p-value method, we find the region to the left of the test statistic. We get 0.00001. We need to get a p-value of $\alpha = 0.05$, so we can reject the null hypothesis.

Solution to Practice Problem 6: *Suppose that we randomly sample 100 manufactured DVDs and find that the mean error rate in storage is 0.1%, with a known population standard deviation of 0.02%. Test that the mean error rate is actually 0.11% at the 0.01 level of significance.*

Our null hypothesis is H_0: $\mu = 0.11$.

Our alternative hypothesis is H_1: $\mu < 0.11$.

The test statistic is found using $n = 100$ and $\bar{x} = 0.10$ with $\sigma = 0.02$. We get $z = \frac{0.10 - 0.11}{0.02/\sqrt{100}} = -5$. The critical value will be $z = -2.575$. Our test statistic falls within the rejection region (it is less than –2.575) so we can reject the null hypothesis. That is, if the mean error rate is 0.11%, we could not have tested 100 of them and obtained 0.10% by chance.

Solution to Practice Problem 7: *Suppose that we randomly sample 20 cups of espresso and find that their mean temperature is 155° degrees, with a sample standard deviation of 8°. Test that the mean temperature of a cup of espresso is actually 165° at the 0.05 level of significance.*

Our null hypothesis is H_0: $\mu = 165$.

Our alternative hypothesis is H_1: $\mu < 165$.

Here, we have a simple random sample with an unknown population standard deviation. So, instead of a z score, we will find a t score. The test statistic is found using $n = 20$ and $\bar{x} = 155$, with $s = 8$ and $d.f. = 20 - 1 = 19$. We get $t = \frac{155-165}{8/\sqrt{20}} = -5.59$.

The significance level is $\alpha = 0.05$ and the critical value will be –1.729 (from the t table). Our test statistic falls within the rejection region (it is less than –1.729, so we can reject the null hypothesis. That is, if the mean temperature of a cup of espresso is 165°, then it is very unlikely that we could have tested 20 of them and obtained 155° by chance.

Solution to Practice Problem 8: *Suppose that we randomly sample the resting pulse rate of 12 female college soccer players and obtain a mean pulse rate of 71 beats per minute, with a standard deviation of 4 beats per minute. Test that the mean pulse rate is actually 75 beats per minute at the 0.01 level of significance.*

Our null hypothesis is H_0: $\mu = 75$.

Our alternative hypothesis is H_1: $\mu < 75$.

Here, we have a simple random sample with an unknown population standard deviation. So, instead of a z score, we will find a t score. The test statistic is found using $n = 12$ and $\bar{x} = 71$ with $\sigma = 4$ and $d.f. = 12 - 1 = 11$. We get $t = \frac{71-75}{4/\sqrt{12}} = -3.46$.

The significance level is $\alpha = 0.05$ and the critical value will be –1.796 (from the t table). Our test statistic falls within the rejection region (it is less than –1.796), so we can reject the null hypothesis. That is, if the mean pulse rate of a female soccer player is 75 beats per minute, then it is very unlikely that we could have tested 12 of them and obtained 71 beats per minute by chance.

Solution to Practice Problem 9: *A freshman at our university usually gains a mean of 7.2 pounds with a standard deviation of 2.1 pounds. A random sample of 21 students obtains a standard deviation of 3.2 pounds. Test the claim that $\sigma = 2.1$ at the 0.05 level of significance.*

Our null hypothesis is H_0: $\sigma = 2.1$.

Our alternative hypothesis is H_1: $\sigma > 2.1$

The significance level is $\alpha = 0.05$ with $21 - 1 = 20$ $d.f.$ We will use the χ^2 distribution to test this claim because it is about a standard deviation.

The test statistic is $\chi^2 = \dfrac{(21-1)(3.2)^2}{(2.1)^2} = 46.4399$.

The critical values are $\chi^2 = 9.591$ and $\chi^2 = 34.170$. Because the test statistic is in the critical value, we can reject the null hypothesis. That is, if the standard deviation were 2.1, then it is highly unlikely that we could have obtained a standard deviation of 3.2 by chance.

Solution to Practice Problem 10: *Suppose that the mean stopping time of a particular model of car moving at 40 mph is listed as 1.8 seconds, with a standard deviation of 0.25 seconds. A random sample of 12 cars obtains a standard deviation of 0.35 seconds. Test the claim that $\sigma = 0.25$ seconds at the 0.05 level of significance.*

Our null hypothesis is H_0: $\sigma = 0.25$.

Our alternative hypothesis is H_1: $\sigma > 0.25$.

The significance level is $\alpha = 0.05$ with $12 - 1 = 11$ *d.f.* We will use the χ^2 distribution to test this claim because it is about a standard deviation.

The test statistic is $\chi^2 = \dfrac{(12-1)(0.35)^2}{(0.25)^2} = 21.56$.

The critical values are $\chi^2 = 7.564$ and $\chi^2 = 30.191$. Because the test statistic is not in the critical value, we *cannot* reject the null hypothesis. That is, if the standard deviation were 0.25, then it is possible that we could have obtained a standard deviation of 0.35 by chance. The best thing to do is to redo the experiment with a larger sample size to see if we still obtain a standard deviation greater than 0.25.

UNIT TWELVE

Working with Two Samples, Correlation, and Regression

Now we have learned 2 important aspects of Statistics–finding confidence intervals and testing hypotheses. Everything that we have done so far involves drawing inferences from 1 sample. Most of the time, we will need to compare 2 sets of data. Sometimes it will be a group that receives a medical treatment versus a group that receives a placebo. Or sometimes we will want to compare what people say about something they do versus what they actually do (for example hours watching television, what they eat, etc.). In most real-world situations, researchers want to test a hypothesis on 2 samples. So now we will learn how to do so. Many of the formulas are very similar to what we have seen for 1 sample, with the second group taken into account. By the way, often the 2 groups are not the same size. This can be because some members have to be disqualified after they have been initially chosen, or because of a significant difference in population size, or other factors.

Suppose that we want to test a hypothesis about 2 population proportions. First, we need to know that the 2 samples are independent. The statistics are the same as we learned before about proportions, but now we have two samples instead of one. That is, we will have p_1 as the population proportion of the first group, and the corresponding $\hat{p}_1$ for the proportion of the first sample. We will also have p_2 as the population proportion of the second group and the corresponding $\hat{p}_2$ as the proportion of the second sample. When we wish to do a hypothesis test on 2 samples, we will test the claim that $p_1 = p_2$, with the alternate hypothesis being that there is a difference between the 2 proportions. Then the z statistic becomes

$$z = \frac{(\hat{p}_1 - \hat{p}_2) - (p_1 - p_2)}{\sqrt{\frac{\overline{p}\overline{q}}{n_1} + \frac{\overline{p}\overline{q}}{n_2}}},$$

where the null hypothesis will have $p_1 - p_2 = 0$, $\hat{p}_1 = \frac{x_1}{n_1}$, $\hat{p}_2 = \frac{x_2}{n_2}$, $\overline{p} = \frac{x_1 + x_2}{n_1 + n_2}$, and $\overline{q} = 1 - \overline{p}$.

This may look intimidating but notice that all we are really doing now is revising the formula to account for the differences between the 2 proportions. The hypothesis test will still follow the same procedure. Let's do an example.

Example 1: Suppose that female freshmen complain that they are unfairly targeted for underage drinking. We randomly surveyed 500 male freshmen and 440 female freshmen who were charged with underage drinking in their freshman year. Of the males, 51 had been charged with underage drinking. Of the females, 48 had been charged with underage drinking. Test the claim that the proportion of female freshmen stopped for underage drinking is greater than the proportion of male freshmen at the $\alpha = 0.05$ level of significance.

Let's summarize our sample data in a table:

Males (Sample 1)	Females (Sample 2)
$n = 500$	$n = 440$
$x_1 = 51$	$x_2 = 48$
$\hat{p}_1 = \frac{x_1}{n_1} = \frac{51}{500} = 0.102$	$\hat{p}_2 = \frac{x_2}{n_2} = \frac{48}{440} = 0.109$

The claim is that women are unfairly charged so the claim is that $p_2 > p_1$. We will then test the following:

H_0: $p_2 = p_1$
H_1: $p_2 > p_1$

at the $\alpha = 0.05$ level of significance.

We have 2 large, independent, random samples, so we can use the normal distribution to test the hypothesis. First, we need to calculate $\overline{p}$. We get $\overline{p} = \frac{51+48}{500+440} = \frac{99}{940} \approx 0.10532$. Note that we are keeping some extra decimal places. This is to avoid rounding errors. If $\overline{p} = 0.10532$, then $\overline{q} = 1 - 0.10532 = 0.89468$. Now we can find the z statistic. We get

$$z = \frac{\left(\frac{48}{440} - \frac{51}{500}\right) - 0}{\sqrt{\frac{(0.10532)(0.89468)}{440} + \frac{(0.10532)(0.89468)}{500}}}$$

$$= \frac{-0.00709}{\sqrt{0.00021415 + 0.00018855}} = 0.353.$$

The calculation is messy so be careful when you do it. Of course, many calculators and computer programs can find the z statistic for you. The area to the left of the z statistic is 0.6368, so the area to the right of it is 1 – 0.6368 = 0.3632. Because this is greater than the significance level of $\alpha = 0.05$, we cannot reject the null hypothesis. That is, based on our data, we do not see enough of a difference in the proportions to support the claim that female freshmen are more likely to be charged with underage drinking than male freshmen are. This does not necessarily mean that the claim is not true, merely that the data are not strong enough to support the claim.

Now we wish to test claims made about data from 2 independent samples. Two samples are *independent* if the sample values from one set are not related to the sample values from the other set. When the sample values are related, they often referred to as *matched pairs*, and the sets are dependent. Suppose that we had 2 groups of test subjects. One is given a new drug to treat kidney infections and the other is given a placebo. The individuals in the first group are not matched in any way with the individuals in the second group, so the samples are independent. Now suppose that we wanted to test the effect of a new low-sodium diet on kidney output. We measured the kidney output of the individuals in the group before and after following the diet. The value before the diet is then *matched* to the value after the diet because the values come from the same individual.

In this case, we will use a t statistic to test a hypothesis because we are sampling from a population. If we want to claim that a treatment is effective, we will be testing a claim about $\mu_1 - \mu_2$. In order to use the t statistic, we will need to know that the samples are independent, that both samples are simple random samples, and that either the sample sizes are both greater than 30 or that they both come from populations that are normally distributed. The test statistic becomes

$$t = \frac{(\bar{x}_1 - \bar{x}_2) - (\mu_1 - \mu_2)}{\sqrt{\frac{s_1^2}{n_1} + \frac{s_2^2}{n_2}}}.$$

The degrees of freedom are found by

$$d.f. = \frac{\left(\frac{s_1^2}{n_1} + \frac{s_2^2}{n_2}\right)^2}{\frac{\left(\frac{s_1^2}{n_1}\right)^2}{n_1 - 1} + \frac{\left(\frac{s_2^2}{n_2}\right)^2}{n_2 - 1}}.$$

As with the proportions, the main difference between these statistics and the 1-sample statistics is that they are adjusted for the fact that there are 2 samples instead of just 1. The calculations are messy, so be careful! Let's do an example.

Example 2: Suppose we wish to test the effectiveness of a new drug that fights kidney infections. We randomly select two groups and administer the drug to one group and a placebo to the other group. We then measure how long the infection lasts in each group (in hours). We obtain the following data:

	Receives Drug	Receives Placebo
n	80	82
$\overline{x}$	34 hours	41 hours
s	5.3	7.6

We wish to test (at the $\alpha = 0.05$ level of significance) the claim that the new drug shortens the length of time of kidney infections. Thus, we have

H_0: $\mu_1 = \mu_2$
H_1: $\mu_1 < \mu_2$

at the $\alpha = 0.05$ level of significance.

The t statistic is

$$t = \frac{(34-41)-0}{\sqrt{\frac{5.3^2}{80}+\frac{7.6^2}{82}}} = -6.8134.$$

The degrees of freedom are:

$$d.f. = \frac{\left(\frac{5.3^2}{80}+\frac{7.6^2}{82}\right)^2}{\frac{\left(\frac{5.3^2}{80}\right)^2}{80-1}+\frac{\left(\frac{7.6^2}{82}\right)^2}{82-1}} = \frac{1.1141}{0.00156-0.00613} \approx 244.$$

If we go to the t table using 120 degrees of freedom (we do not have 240) and an area of 0.05, we get $t = -1.653$. Our t statistic falls within the critical region, so we can reject the null hypothesis and conclude that our drug does reduce the length of time of a kidney infection.

What if we are working with 2 samples that are not independent? In other words, what if the values of 1 population are somehow related to the values from the other population? As we said earlier, these are often called *matched pairs*. In order to conduct a hypothesis test, we will assume that the sample consists of matched pairs, that the samples are simple random samples, and that the sample is large ($n > 30$)

or that the sample distribution is approximately normal. We will use the following statistics:

Let d be the individual difference between the two values of a matched pair.

Let μ_d be the mean value of the differences, d, for the population of the matched pairs.

Let $\bar{d}$ be the mean value of the differences, d, for the matched pairs in the sample.

Let s_d be the standard deviation of the differences, d, for the matched pairs in the sample.

Let n be the number of *pairs* in the sample.

Then the hypothesis test statistic is: $t = \dfrac{\bar{d} - \mu_d}{s_d / \sqrt{n}}$. The degrees of freedom is $d.f. = n - 1$.

Let's do an example.

Example 3: Suppose that we develop a computer model that will estimate the monthly rainfall for our city. We want to compare it to the actual rainfall to see how accurate our estimates are. We have the following data (in inches) for the past year:

Actual Rainfall	4.3	4.1	3.8	4.4	4.9	2.7	3.6	4.5	4.2	4.4	4.1	3.9
Estimate	4.7	4.2	4.4	4.0	4.2	4.1	3.9	4.0	4.1	3.9	3.9	4.1
Difference	−0.4	−0.1	−0.6	0.4	0.7	−1.4	−0.3	0.5	0.1	0.5	0.2	−0.2

We want to know if there is sufficient evidence to claim that the difference between our model and the actual rainfall is not 0. We wish to test our claim at the $\alpha = 0.05$ level of significance.

We have

H_0: $\mu_d = 0$

H_1: $\mu_d \neq 0$.

at the $\alpha = 0.05$ level of significance.

First, we will need to find $\bar{d}$, the mean value of the differences. We get $\bar{d} = -0.05$. Next, we will need to find s_d, the standard deviation of the differences. We get $s_d = 0.587$. Our degrees of freedom are $d.f. = 12 - 1 = 11$. Now we can find the test statistic $t = \dfrac{-0.05 - 0}{0.587 / \sqrt{12}} = -0.295$. From the t table, we get $t_{0.05} = 2.201$. The test statistic does not fall in the critical region, so we *cannot* reject the null hypothesis. That is, the data do not provide sufficient evidence that there is a difference between our estimates of monthly rainfall and the actual monthly rainfall.

Now that we have looked at matched pairs, we will look at whether a relationship exists between 2 populations. Such a relationship between 2 variables is called a *correlation*. In this book, we will only look at linear relationships, although many other kind of relationships exist.

Often, we can see a relationship between two variables by graphing each paired sample with a dot on a pair of coordinate axis. The resulting plot is called a *scatterplot*. When we look at a scatterplot, we can sometimes see a pattern that indicates a possible correlation between 2 variables. The following are different types of scatterplots that we could get.

Figure 26

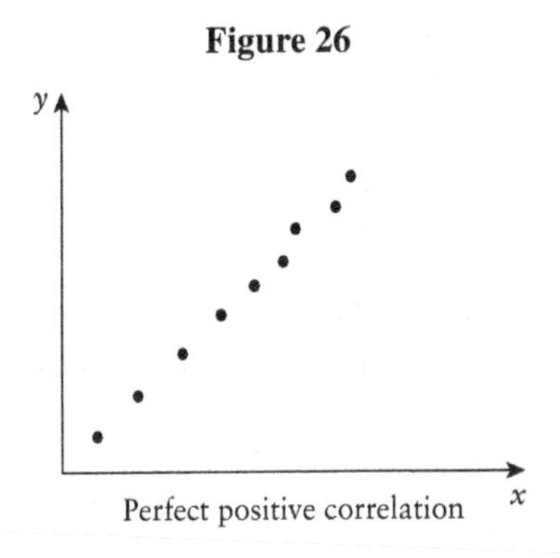

Perfect positive correlation

In the above plot, there is a perfect positive correlation between x and y. Note how the dots form a straight line with a positive slope. Such a correlation is rare.

Figure 27

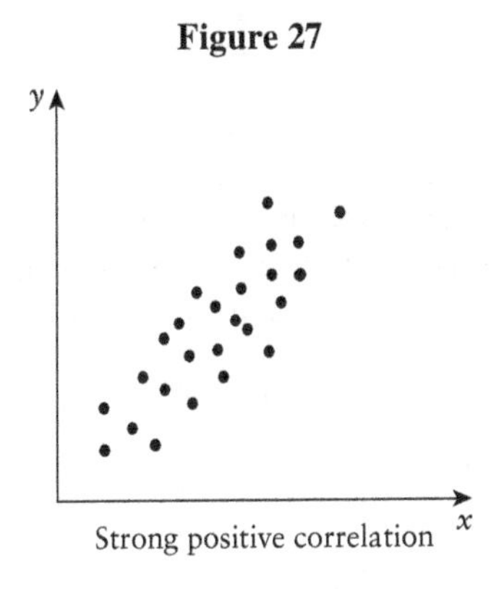

Strong positive correlation

In the above plot, there is a strong positive correlation between x and y. Note how the dots seem to be tightly bunched around a straight line with a positive slope.

Figure 28

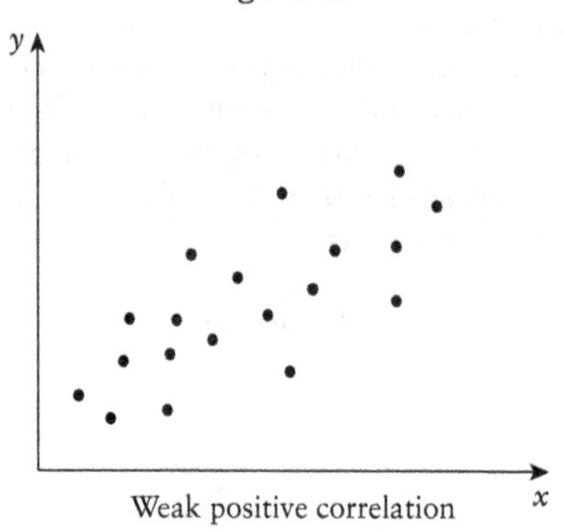

In this plot, there is a positive correlation between x and y. Note how the dots seem to be loosely bunched around a straight line with a positive slope.

Figure 29

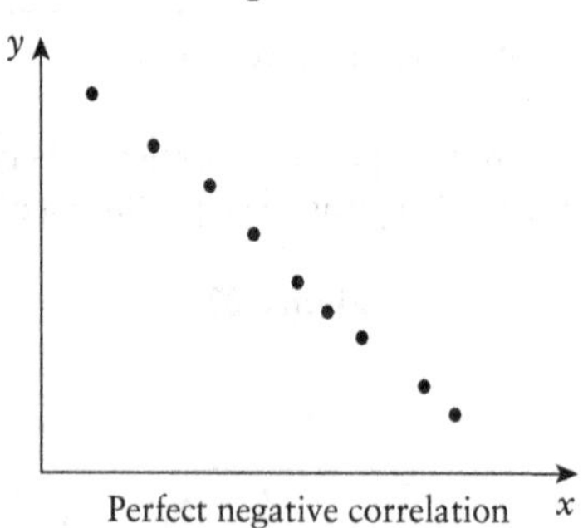

In the above plot, there is a perfect negative correlation between x and y. Note how the dots form a straight, line with a negative slope. Such a correlation is rare.

Figure 30

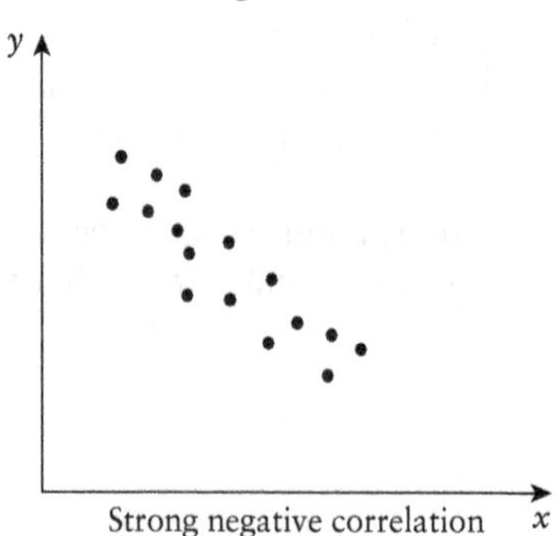

In the above plot, there is a strong negative correlation between x and y. Note how the dots seem to be tightly bunched around a straight line with a negative slope.

Figure 31

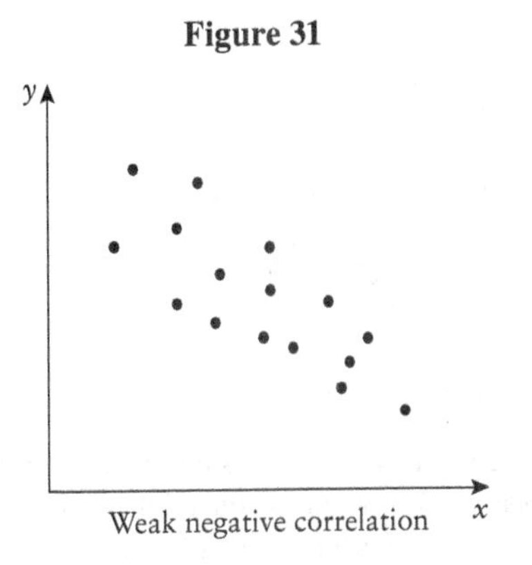

Weak negative correlation

In the above plot, there is a negative correlation between x and y. Note how the dots seem to be loosely bunched around a straight line with a negative slope.

In the six previous plots, the correlations vary from perfect to strong to weak.

Figure 32

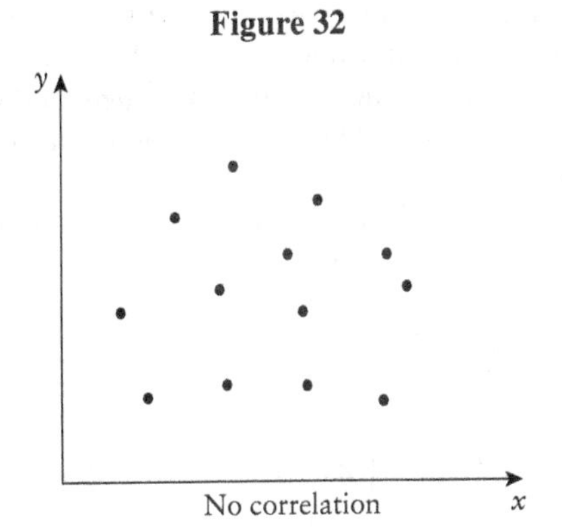

No correlation

It is also possible to have no correlation, as shown above.

Figure 33

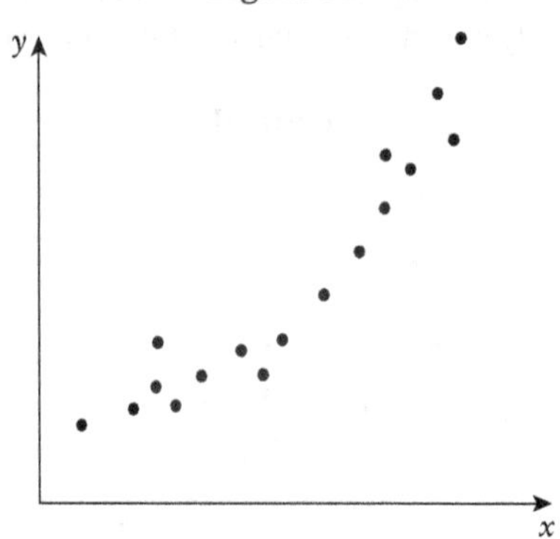

It is also possible to have a correlation that is not linear. For example, the dots in the scatterplot above appear to follow an exponential curve. As we stated previously, such correlations are beyond the scope of this book.

Of course, we cannot just rely on a visual guess if 2 variables are correlated. We will need a way to measure the relationship between the 2 variables to see if they are correlated and, if so, how strong the correlation is. We do so using the *linear correlation coefficient* (also known as the *Pearson product moment correlation coefficient*). The coefficient, r, looks at a random sample of pairs of data (x, y), which are drawn from a random sample of data. The pairs of data must have a *bivariate normal distribution*, which is too complicated for this book but suffice it to say that both the x and y distributions must be approximately bell-shaped, that is, normally distributed.

The linear correlation coefficient is calculated according to the following formula:

$$r = \frac{n\sum_{i=1}^{n} x_i y_i - \sum_{i=1}^{n} x_i \sum_{i=1}^{n} y_i}{\sqrt{n\sum_{i=1}^{n} x_i^2 - \left(\sum_{i=1}^{n} x_i\right)^2}\sqrt{n\sum_{i=1}^{n} y_i^2 - \left(\sum_{i=1}^{n} y_i\right)^2}}.$$

When calculating r, be careful not to confuse $\sum_{i=1}^{n} x_i^2$ with $\left(\sum_{i=1}^{n} x_i\right)^2$. In the former, we first square the x values and then sum them; in the latter, we first sum the x values and then square the sum. These are not the same. To convince yourself, try an easy example with the numbers 1, 2, and 3. The sum of the squares is $\sum_{i=1}^{3} x_i^2 = 1^2 + 2^2 + 3^2 = 1 + 4 + 9 = 14$, whereas the square of the sum is $\left(\sum_{i=1}^{3} x_i\right)^2 = (1+2+3)^2 = 6^2 = 36$. Let's find the linear correlation coefficient for a simple example.

Example 4: Suppose that we have the following set of data. Find the linear correlation coefficient of the data set.

x	y
5	7
9	8
12	8
17	11
2	4
6	6
8	8
3	5
19	10
16	11

Let's calculate each component of the formula.

x	y	xy	x^2	y^2
5	7	35	25	49
9	8	72	81	64
12	8	96	144	64
17	11	187	289	121
2	4	8	4	16
6	6	36	36	36
8	8	64	64	64
3	5	15	9	25
19	10	190	361	100
16	11	176	256	121
$\sum_{i=1}^{10} x_i = 97$	$\sum_{i=1}^{10} y_i = 78$	$\sum_{i=1}^{10} x_i y_i = 879$	$\sum_{i=1}^{10} x_i^2 = 1269$	$\sum_{i=1}^{10} y_i^2 = 660$

Now let's plug the information into the formula

$$r = \frac{n\sum_{i=1}^{n} x_i y_i - \sum_{i=1}^{n} x_i \sum_{i=1}^{n} y_i}{\sqrt{n\sum_{i=1}^{n} x_i^2 - \left(\sum_{i=1}^{n} x_i\right)^2}\sqrt{n\sum_{i=1}^{n} y_i^2 - \left(\sum_{i=1}^{n} y_i\right)^2}}.$$

We get

$$r = \frac{(10)(879)-(97)(78)}{\sqrt{(10)(1269)-(97)^2}\sqrt{(10)(660)-(78)^2}} = \frac{1224}{\sqrt{3281}\sqrt{516}} = 0.9407.$$

Now that we have found r, what does it mean? The value of r must always be $-1 \le r \le 1$. If r is close to 0, then we can conclude that there is no significant linear correlation between the two variables. The closer r is to 1, the stronger the positive linear correlation is. The closer r is to -1, the stronger the negative linear correlation is. Of course, our conclusion rests on the quality of our sample of data. Also, the term "close" is not well defined. If we look at Table 4 in the back of the book, we can use the following rule: If the absolute value of r is greater than the value in the table, then there is a significant linear correlation between the 2 variables. Otherwise, the evidence does not support the conclusion that there is a significant linear correlation between the 2 variables. In this case, with $n = 10$ and $r = 0.9407$, we can conclude that there is a significant linear correlation between x and y.

As we almost always find in Statistics, note that we cannot draw definite conclusions. Instead, we say things like "There is evidence of a strong relationship." or "We cannot say that there is evidence of a strong relationship." This is one of the most frustrating aspects of Statistics for the average person who is usually seeking definite answers.

When we have a linear correlation between 2 variables, then we can construct an equation that relates x and y. This equation can then be used to predict a value of y for a given value of x. However, the predicted value of y is not the same as the actual value of y. We always have to be alert to the possibility that there are other factors that can influence the relationship between x and y (among others, random variation). We can find this with r^2, which is the proportion of the variation in y that is due to the linear relationship between x and y. In the above example, $r^2 = 0.8849$, which means that about 88% of the value of y is due to the linear relationship between x and y. Of course, this means that about 12% of the value of y is due to other factors. There is also something very important to bear in mind when looking at correlation. When we learn that one variable is correlated with another, we found that the 2 variables are related to each other in one way. We have *not* found whether one variable *causes* the other variable to respond. This is best summed up on a phrase that is widely known in Statistics: Correlation does not imply causation. Remember this when you read about the outcome of an experiment, particularly in fields like medicine or the social sciences. Just because 2 variables occur together does *not* mean that one causes the other. There could be, and often is, at least 1 other variable that could be the true cause of the relationship between the 2 variables. This is the difficult and complex work that Statisticians do every day and is far beyond the scope of this book. For now, just remember: Correlation does not imply causation.

Now that we have learned about correlation and how it tells us if there is a relationship between two variables, let's learn about *regression*, which helps us predict the value of one of the variables, given a value of the other variable. We will do this by finding a *regression line*, which describes a linear relationship between two variables through a *regression equation.* There are many types of regression – linear, exponential, logarithmic, quadratic, etc.–but in this book we will only look at linear regression.

The regression equation gives us a relationship between x, which is the independent variable (sometimes called the *predictor* or *explanatory* variable) and y, which is the dependent variable (sometimes called the *response* variable). A typical regression is of the form $y = ax + b$. We can find the slope, a, and the y-intercept, b, by the following equations:

$$a = \frac{n\left(\sum_{i=1}^{n} x_i y_i\right) - \left(\sum_{i=1}^{n} x_i\right)\left(\sum_{i=1}^{n} y_i\right)}{n\left(\sum_{i=1}^{n} x_i^2\right) - \left(\sum_{i=1}^{n} x_i\right)^2}$$

and $b = \overline{y} - a\overline{x}$.

Example 5: Find the linear regression equation of the data set from Example 4.

x	y
5	7
9	8
12	8
17	11
2	4
6	6
8	8
3	5
19	10
16	11

We already have the components of the regression equation from the example:

x	y	xy	x^2	y^2
5	7	35	25	49
9	8	72	81	64
12	8	96	144	64
17	11	187	289	121
2	4	8	4	16

6	6	36	36	36
8	8	64	64	64
3	5	15	9	25
19	10	190	361	100
16	11	176	256	121
$\sum_{i=1}^{10} x_i = 97$	$\sum_{i=1}^{10} y_i = 78$	$\sum_{i=1}^{10} x_i y_i = 879$	$\sum_{i=1}^{10} x_i^2 = 1269$	$\sum_{i=1}^{10} y_i^2 = 660$

Now we can plug these into the formula for the slope. We get $a = \frac{10(879) - (97)(78)}{10(1269) - (97)^2} = \frac{1224}{3281} = 0.3731.$

Now we need to find $\overline{x}$ and $\overline{y}$. They are $\overline{x} = \frac{97}{10} = 9.7$ and $\overline{y} = \frac{78}{10} = 7.8$. We can plug these into the equation for the y-intercept. We get $b = 7.8 - (0.3731)(9.7) = 4.1809$.

Therefore, the regression equation is $y = 0.3731x + 4.1809$.

These days, calculators and computer programs can calculate regression equations and correlation coefficients (along with many other statistics), so there is really no reason to calculate these by hand. Nonetheless, it is important to know how to find the statistics.

Now that we have a regression equation, we can use it to predict the value of a variable. First, let's test the equation that we just found at $x = 12$ and $x = 5$. If we plug $x = 12$ into the regression equation, we get $y = 0.3731(12) + 4.1809 = 8.658$, and if we plug $x = 5$ into the regression equation, we get $y = 0.3731(5) + 4.1809 = 6.0464$. The actual value for $x = 12$ is $y = 8$ and for $x = 5$, $y = 7$. Our regression equation looks pretty good. Now let's predict the value of y when $x = 25$. If we plug $x = 25$ into the regression equation, we get $y = 0.3731(25) + 4.1809 = 13.5084$.

When we use a regression equation to make predictions, it is important to realize that the farther the x is from the known data set, the less accurate the prediction might be. In other words, in our example, the predicted value of y when $x = 100$ might be a less accurate prediction than the predicted y for $x = 25$.

There are other factors to bear in mind when using a regression equation. First, and foremost, the relationship between the 2 variables may not be linear. Second, the regression equation can be influenced by outliers or by points that may strongly influence the regression equation. We can usually identify these latter points by looking at how far these points are away *horizontally* from the other points. We can also look at the *residuals* of the regression equation. A *residual* is the difference between the observed value of y and the predicted value of y. A regression line is a *least-squares* line if the sum of the squares of the residuals is the smallest possible sum. We usually find the residuals and the least-squares line with a calculator or a computer program. As you can imagine, there is much more to be done with 2 samples and with multiple samples, but most of it is beyond the scope of this book. This chapter really serves to introduce the subject, and, if the reader wishes to explore these subjects in depth, there are many excellent books.

When you take statistics, you will usually not be asked to find any of the calculations in this chapter by hand. The calculations are tedious and time-consuming, and can be found so much more easily using a calculator, a spreadsheet, or a statistics program. There are even websites where you can enter your data and it will give you the statistics that you need. The purpose of this chapter, and, in fact, of this book, is to help you understand how to calculate the statistics and how to interpret them. In the meantime, we hope that with this book, you will now be comfortable with analyzing and measuring the data for 1 sample, testing a hypothesis using that data, and working with more than 1 sample of data.

Are you ready to practice?

Practice Problems

Practice Problem 1: Suppose that we wish to determine whether men in our county are more likely to be charged with a felony when arrested for DWI (Driving While Intoxicated) than women are. We randomly surveyed 300 men and 280 women of those who were arrested for DWI last year. Of the men, 212 had been charged with a felony. Of the females, 188 had been charged with a felony. Test the claim that the proportion of men charged with a felony for DWI is greater than the proportion of women at the $\alpha = 0.05$ level of significance.

Practice Problem 2: The local animal shelter takes in stray puppies and kittens and usually can find a family to adopt them within a week. They wish to determine if puppies are more likely to be adopted within a week than kittens are. We randomly selected 170 of the puppies and 195 of the kittens who were adopted last year. Of the puppies, 145 were adopted within a week. Of the kittens, 151 were adopted within a week. Test the claim at the $\alpha = 0.05$ level of significance that the proportion of puppies adopted within a week is greater than the proportion of kittens.

Practice Problem 3: Suppose we wish to test the response time of a new antacid that reduces acid reflux. We randomly select 2 groups and administer the antacid to one group and a placebo to the other group. We then measure how quickly the stomach acid returns to a normal pH level (in minutes). We obtain the following data:

	Receives Antacid	**Receives Placebo**
n	60	66
$\overline{x}$	11.2	14.1
s	2.1	2.4

We wish to test (at the $\alpha = 0.05$ level of significance) the claim that the new antacid reduces the response time to eliminate acid reflux.

Practice Problem 4: We create a new abdominal workout and we want to test whether it is more effective than the old one that we were teaching. We create 2 large groups–one does the new workout for 6 weeks and the other continues with the old workout. After 6 weeks, we randomly select 2 sub-groups from the 2 original groups and measure how many sit-ups each group can perform in 60 seconds. We obtain the following data:

	New Workout	**Old Workout**
n	25	24
$\overline{x}$	49	46
s	1.6	1.4

We wish to test (at the $\alpha = 0.05$ level of significance) the claim that the new workout increases the number of sit-ups that a person can perform in 60 seconds.

Practice Problem 5: We develop a new computer model that estimates traffic wait time for commuters at the local tollbooth and we wish to compare it to the actual wait times to see how accurate our estimates are. We have the following average wait times (in minutes) for the past 10 days:

Actual Wait Time	1.8	2.6	3.8	1.4	5.8	1.9	4.6	9.5	2.2	3.1
Estimate	2.3	3.6	4.4	1.0	3.2	3.1	4.5	8.1	4.1	3.8
Difference	−0.5	−1.0	−0.6	0.4	2.6	−1.2	0.1	1.4	−1.9	−0.7

We want to know if there is sufficient evidence to claim that the difference between our model and the actual wait time is not 0. We wish to test our claim at the $\alpha = 0.05$ level of significance.

Practice Problem 6: We have developed a new set of low-calorie meals and wish to see if the meals are effective in reducing weight. The subjects only eat our new meals for a 2-week period. We have the following "before" and "after" weights (in pounds) for 11 men:

Before	188	206	195	293	267	303	198	227	244	251	317
After	180	211	191	281	265	290	199	230	244	241	308
Difference	8	−5	4	12	2	13	−1	−3	0	10	9

We want to know if there is sufficient evidence to claim that our new low-calorie meals will reduce a person's weight in just 2 weeks. We wish to test our claim at the $\alpha = 0.05$ level of significance.

Practice Problem 7: Suppose that we have the following set of data of weights (in pounds) and resting pulse rates for a randomly chosen sample of 50-year-old men. Find the linear correlation coefficient of the data set.

Weight	Pulse Rate
195	75
188	74
150	68
220	86
205	88
152	66
189	80
275	95
215	90
163	78

Practice Problem 8: Suppose that we have the following set of data of Grade Point Averages (out of a possible 4 points) and the number of hours a day spent watching television. Find the linear correlation coefficient of the data set.

G.P.A	Hours of TV
3.92	1.2
3.78	2.8
2.23	4.5
2.74	3.1
3.31	3.5
3.82	3.1
3.02	2.8
3.17	4.1
2.88	6.2
2.96	1.9
3.55	1.3
4.0	2.1
3.67	0
3.31	1.6
3.33	1.8

Practice Problem 9: Suppose that we have the following set of data of Grade Point Averages (out of a possible 4 points) from Practice Problem 8 and the number of hours a day spent listening to the radio. Find the Linear Correlation Coefficient of the data set.

G.P.A	Hours of Radio
3.92	4.2
3.78	2.8
2.23	1.5
2.74	1.7
3.31	3.5
3.82	4.2
3.02	2.9
3.17	4.1
2.88	3.6
2.96	4.1
3.55	2.7
4.0	2.8
3.67	3.3
3.31	1.6
3.33	5.8

Practice Problem 10: Find the linear regression equation of the data set from Practice Problem 7.

Weight	Pulse Rate
195	75
188	74
150	68
220	86
205	88
152	66
189	80
275	95
215	90
163	78

Practice Problem 11: Find the Linear Regression equation of the data set from Practice Problem 8

G.P.A	Hours of TV
3.92	1.2
3.78	2.8
2.23	4.5
2.74	3.1
3.31	3.5
3.82	3.1
3.02	2.8
3.17	4.1
2.88	6.2
2.96	1.9
3.55	1.3
4.0	2.1
3.67	0
3.31	1.6
3.33	1.8

Practice Problem 12: Find the Linear Regression equation of the data set from Practice Problem 9.

G.P.A	Hours of Radio
3.92	4.2
3.78	2.8
2.23	1.5
2.74	1.7
3.31	3.5
3.82	4.2
3.02	2.9
3.17	4.1
2.88	3.6
2.96	4.1
3.55	2.7
4.0	2.8
3.67	3.3
3.31	1.6
3.33	5.8

Solutions to Practice Problems

Solution to Practice Problem 1: *Suppose that we wish to determine whether men in our county are more likely to be charged with a felony when arrested for DWI (Driving While Intoxicated) than women are. We randomly surveyed 300 men and 280 women of those who were arrested for DWI last year. Of the men, 212 had been charged with a felony. Of the females, 188 had been charged with a felony. Test the claim that the proportion of men charged with a felony for DWI is greater than the proportion of women at the $\alpha = 0.05$ level of significance.*

Let's summarize our sample data in a table:

Men (Sample 1)	Women (Sample 2)
$n = 300$	$n = 280$
$x_1 = 212$	$x_2 = 188$
$\hat{p}_1 = \frac{x_1}{n_1} = \frac{212}{300} = 0.707$	$\hat{p}_2 = \frac{x_2}{n_2} = \frac{188}{280} = 0.671$

The claim is that men are unfairly charged so the claim is that $p_1 > p_2$. We will then test the following:

H_0: $p_1 = p_2$

H_1: $p_1 > p_2$

at the $\alpha = 0.05$ level of significance.

We have 2 large, independent, random samples, so we can use the normal distribution to test the hypothesis. First, we need to calculate $\overline{p}$. We get $\overline{p} = \frac{212+188}{300+280} = \frac{400}{580} \approx 0.6897$. Note that we are keeping some extra decimal places. This is to avoid rounding errors. If $\overline{p} = 0.6897$, then $\overline{q} = 1 - 0.6897 = 0.3103$. Now we can find the z statistic. We get

$$z = \frac{\left(\frac{212}{300} - \frac{188}{280}\right) - 0}{\sqrt{\frac{(0.6897)(0.3103)}{300} + \frac{(0.6897)(0.3103)}{280}}} = \frac{0.03523}{\sqrt{0.0007134 + 0.0007643}} = 0.9165.$$

The area to the left of the z statistic is 0.8212 so the area to the right of it is $1 - 0.8212 = 0.1788$. Because this is greater than the significance level of $\alpha = 0.05$, we cannot reject the null hypothesis. That is, based on our data, we do not see enough of a difference in the proportions to support the claim that men are more likely to be charged with a felony when arrested for DWI than women are. This does not necessarily mean that the claim is not true, merely that the data are not strong enough to support the claim.

Solution to Practice Problem 2: *The local animal shelter takes in stray puppies and kittens and usually can find a family to adopt them within a week. They wish to determine if puppies are more likely to be adopted within a week than kittens are. We randomly selected 170 of the puppies and 195 of the kittens who were adopted last year. Of the puppies, 145 were adopted within a week. Of the kittens, 151 were adopted within a week. Test the claim at the $\alpha = 0.05$ level of significance that the proportion of puppies adopted within a week is greater than the proportion of kittens.*

Let's summarize our sample data in a table:

Puppies (Sample 1)	Kittens (Sample 2)
$n = 170$	$n = 195$
$x_1 = 145$	$x_2 = 151$
$\hat{p}_1 = \frac{x_1}{n_1} = \frac{145}{170} = 0.8529$	$\hat{p}_2 = \frac{x_2}{n_2} = \frac{151}{195} = 0.7744$

The claim is that puppies are more likely to be adopted within a week so the claim is that $p_1 > p_2$. We will then test the following:

H_0: $p_1 = p_2$
H_1: $p_1 > p_2$

at the $\alpha = 0.05$ level of significance.

We have 2 large, independent, random samples, so we can use the normal distribution to test the hypothesis. First, we need to calculate $\overline{p}$. We get $\overline{p} = \frac{145+151}{170+195} = \frac{296}{365} \approx 0.8110$. If $\overline{p} = 0.8110$, then $\overline{q} = 1 - 0.8110 = 0.1890$. Now we can find the z statistic. We get

$$z = \frac{\left(\frac{145}{170} - \frac{151}{195}\right) - 0}{\sqrt{\frac{(0.8110)(0.1890)}{170} + \frac{(0.8110)(0.1890)}{195}}} = \frac{0.0786}{\sqrt{0.0009016 + 0.0007860}} = 1.913.$$

The area to the left of the z statistic is 0.9713, so the area to the right of it is $1 - 0.9713 = 0.0287$. Because this is less than the significance level of $\alpha = 0.05$, we can reject the null hypothesis. That is, based on our data, we do see enough of a difference in the proportions to support the claim that puppies are more likely to be adopted within a week than kittens are. This does not necessarily mean that the claim is not true, merely that the data support the claim.

Solution to Practice Problem 3: *Suppose we wish to test the response time of a new antacid that reduces acid reflux. We randomly select 2 groups and administer the antacid to one group and a placebo to the other group. We then measure how* *quickly the stomach acid returns to a normal pH level (in minutes). We obtain the following data:*

	Receives Antacid	***Receives Placebo***
n	*60*	*66*
$\bar{x}$	*11.2*	*14.1*
s	*2.1*	*2.4*

We wish to test (at the $\alpha = 0.05$ level of significance) the claim that the new antacid reduces the response time to eliminate acid reflux.

Either the antacid reduces the response time significantly faster than the placebo or there is no difference in response times.

Thus, we test:

H_0: $\mu_1 = \mu_2$

H_1: $\mu_1 < \mu_2$

at the $\alpha = 0.05$ level of significance.

The t statistic is: $t = \dfrac{(11.2-14.1)-0}{\sqrt{\dfrac{2.1^2}{60}+\dfrac{2.4^2}{66}}} = -7.2326$. The degrees of freedom are:

$$d.f. = \frac{\left(\dfrac{2.1^2}{60}+\dfrac{2.4^2}{66}\right)^2}{\dfrac{\left(\dfrac{2.1^2}{60}\right)^2}{60-1}+\dfrac{\left(\dfrac{2.4^2}{66}\right)^2}{66-1}} = \frac{0.0258}{0.0000916+0.0001718} \approx 97.95.$$

If we go to the t table, using 60 degrees of freedom (we do not have 98) and an area of 0.05, we get $t = -2.00$. Our t statistic falls within the critical region, so we can reject the null hypothesis and conclude that our new antacid does reduce the length of time to restore a normal pH.

Solution to Practice Problem 4: *We create a new abdominal workout and we want to test whether it is more effective than the old one that we were teaching. We create 2 groups–one does the new workout for 6 weeks and the other continues with the old workout. After 6 weeks, we randomly select 2 groups and measure how many sit ups each group can perform in 60 seconds. We obtain the following data:*

	New workout	***Old Workout***
n	*25*	*24*
$\bar{x}$	*49*	*46*
s	*1.6*	*1.4*

We wish to test (at the α = 0.05 level of significance) the claim that the new workout increases the number of sit-ups that a person can perform in 60 seconds.

The question is whether the new workout results in being able to do more sit -ups than the old workout does, or there is no difference in response times.

Thus, we test:

H_0: $\mu_1 = \mu_2$

H_1: $\mu_1 > \mu_2$

at the $\alpha = 0.05$ level of significance.

The t statistic is $t = \dfrac{(49-46)-0}{\sqrt{\dfrac{1.6^2}{25}+\dfrac{1.4^2}{24}}} = 6.9925$. The degrees of freedom are:

$$d.f. = \frac{\left(\dfrac{1.6^2}{25}+\dfrac{1.4^2}{24}\right)^2}{\dfrac{\left(\dfrac{1.6^2}{25}\right)^2}{25-1}+\dfrac{\left(\dfrac{1.4^2}{24}\right)^2}{24-1}} = \frac{0.03388}{0.0004369+0.00028998} \approx 46.61.$$

If we go to the t table, using 40 degrees of freedom (we do not have 47) and an area of 0.05, we get t = 2.021. Our t statistic falls within the critical region, so we can reject the null hypothesis and conclude that our new workout increases the number of sit-ups that a person can do in 60 seconds.

Solution to Practice Problem 5: *We develop a new computer model that estimates traffic wait time for commuters at the local tollbooth, and we wish to compare it to the actual wait times to see how accurate our estimates are. We have the following average wait time (in minutes) for the past 10 days:*

Actual Wait Time	*1.8*	*2.6*	*3.8*	*1.4*	*5.8*	*1.9*	*4.6*	*9.5*	*2.2*	*3.1*
Estimate	*2.3*	*3.6*	*4.4*	*1.0*	*3.2*	*3.1*	*4.5*	*8.1*	*4.1*	*3.8*
Difference	*−0.5*	*−1.0*	*−0.6*	*0.4*	*2.6*	*−1.2*	*0.1*	*1.4*	*−1.9*	*−0.7*

We want to know if there is sufficient evidence to claim that the difference between our model and the actual wait time is not zero. We wish to test our claim at the α = 0.05 level of significance.

We test:

H_0: $\mu_d = 0$

H_1: $\mu_d \neq 0$

at the $\alpha = 0.05$ level of significance.

First, we will need to find $\bar{d}$ the mean value of the differences. We get $\bar{d} = -0.14$. Next, we will need to find s_d, the standard deviation of the differences. We get $s_d = 1.3268$. Our degrees of freedom are $d.f. = 10 - 1 = 9$. Now we can find the test statistic $t = \frac{-0.14 - 0}{1.3268/\sqrt{10}} = -0.3337$. From the t table, we get $t_{0.05} = -2.262$. The test statistic does not fall in the critical region, so we *cannot* reject the Null Hypothesis. That is, the data do not provide sufficient evidence that there is a difference between our estimates of wait times and the actual wait times.

Solution to Practice Problem 6: *We have developed a new set of low-calorie meals and wish to see if the meals are effective in reducing weight. The subjects only eat our new meals for a 2-week period. We have the following "before" and "after" weights (in pounds) for 11 men:*

Before	188	206	195	293	267	303	198	227	244	251	317
After	180	211	191	281	265	290	199	230	244	241	308
Difference	8	–5	4	12	2	13	–1	–3	0	10	9

We want to know if there is sufficient evidence to claim that our new low-calorie meals will reduce a person's weight in just 2 weeks. We wish to test our claim at the $\alpha = 0.05$ level of significance.

We test:

$H_0: \mu_d = 0$

$H_1: \mu_d \neq 0$

at the $\alpha = 0.05$ level of significance.

First, we will need to find $\bar{d}$, the mean value of the differences. We get $\bar{d} = 4.455$. Next, we will need to find s_d, the standard deviation of the differences. We get $s_d = 6.283$. Our degrees of freedom are $d.f. = 11 - 1 = 10$. Now we can find the test statistic $t = \frac{4.455 - 0}{6.283/\sqrt{11}} = 2.352$. From the t table, we get $t_{0.05} = 2.228$. The test statistic falls within the critical region, so we can reject the null hypothesis. That is, the data provide sufficient evidence that eating our low-calorie meals helps a person to lose weight.

Solution to Practice Problem 7: *Suppose that we have the following set of data of weights (in pounds) and resting pulse rates for a randomly chosen sample of 50-year-old men. Find the linear correlation coefficient of the data set.*

Weight	***Pulse Rate***
195	*75*
188	*74*

150	*68*
220	*86*
205	*88*
152	*66*
189	*80*
275	*95*
215	*90*
163	*78*

Let's calculate each component of the formula. Let x = weight and y = pulse rate.

x	y	xy	x^2	y^2
195	75	14,625	38,025	5625
188	74	13,912	35,344	5476
150	68	10,200	22,500	4624
220	86	18,920	48,400	7396
205	88	18,040	42,025	7744
152	66	10,032	23,104	4356
189	80	15,120	35,721	6400
275	95	26,125	75,625	9025
215	90	19,350	46,225	8100
163	78	12,714	26,569	6084
$\sum_{i=1}^{10} x_i = 1952$	$\sum_{i=1}^{10} y_i = 800$	$\sum_{i=1}^{10} x_i y_i = 159{,}038$	$\sum_{i=1}^{10} x_i^2 = 393{,}538$	$\sum_{i=1}^{10} y_i^2 = 64{,}830$

Now let's plug the information into the formula

$$r = \frac{n\sum_{i=1}^{n} x_i y_i - \sum_{i=1}^{n} x_i \sum_{i=1}^{n} y_i}{\sqrt{n\sum_{i=1}^{n} x_i^2 - \left(\sum_{i=1}^{n} x_i\right)^2}\sqrt{n\sum_{i=1}^{n} y_i^2 - \left(\sum_{i=1}^{n} y_i\right)^2}}.$$

We get

$$r = \frac{(10)(159{,}038) - (1952)(800)}{\sqrt{(10)(393{,}538) - (1952)^2}\sqrt{(10)(64{,}830) - (800)^2}} = \frac{28{,}780}{\sqrt{125{,}076}\sqrt{8300}} = 0.8932.$$

If we look at the table in the back of the book, we can conclude that, with $n = 10$ and $r = 0.8932$, there is a significant linear correlation between x and y.

Solution to Practice Problem 8: *Suppose that we have the following set of data of Grade Point Averages (out of a possible 4 points) and the number of hours a day spent watching television. Find the linear correlation coefficient of the data set.*

G.P.A	Hours of TV
3.92	1.2
3.78	2.8
2.23	4.5
2.74	3.1
3.31	3.5
3.82	3.1
3.02	2.8
3.17	4.1
2.88	6.2
2.96	1.9
3.55	1.3
4.0	2.1
3.67	0
3.31	1.6
3.33	1.8

Let's calculate each component of the formula. Let x = G.P.A. and y = Hours of TV.

x	y	xy	x^2	y^2
3.92	1.2	4.704	15.366	1.44
3.78	2.8	10.584	14.288	7.84
2.23	4.5	10.035	4.9729	20.25
2.74	3.1	8.494	7.5076	9.61
3.31	3.5	11.585	10.956	12.25
3.82	3.1	11.842	14.592	9.61
3.02	2.8	8.456	9.1204	7.84
3.17	4.1	12.997	10.049	16.81
2.88	6.2	17.856	8.2944	38.44
2.96	1.9	5.624	8.7616	3.61
3.55	1.3	4.615	12.603	1.69
4.0	2.1	8.4	16	4.41

3.67	0	0	13.469	0
3.31	1.6	5.296	10.956	2.56
3.33	1.8	5.994	11.089	3.24
$\sum_{i=1}^{15} x_i = 49.69$	$\sum_{i=1}^{15} y_i = 40$	$\sum_{i=1}^{15} x_i y_i = 126.482$	$\sum_{i=1}^{15} x_i^2 = 168.0255$	$\sum_{i=1}^{15} y_i^2 = 139.6$

Now let's plug the information into the formula:

$$r = \frac{(15)(126.482) - (49.69)(40)}{\sqrt{(15)(168.0255) - (49.69)^2}\sqrt{(15)(139.6) - (40)^2}} = \frac{-90.37}{\sqrt{51.2864}\sqrt{494}} = -0.5678.$$

If we look at the table in the back of the book, we can conclude that, with $n = 15$ and $r = -0.5678$, there is a significant negative linear correlation between x and y.

Solution to Practice Problem 9: *Suppose that we have the following set of data of Grade Point Averages (out of a possible 4 points) from Practice Problem 8 and the number of hours a day spent listening to the radio. Find the Linear Correlation Coefficient of the data set.*

G.P.A	*Hours of Radio*
3.92	*4.2*
3.78	*2.8*
2.23	*1.5*
2.74	*1.7*
3.31	*3.5*
3.82	*4.2*
3.02	*2.9*
3.17	*4.1*
2.88	*3.6*
2.96	*4.1*
3.55	*2.7*
4.0	*2.8*
3.67	*3.3*
3.31	*1.6*
3.33	*5.8*

Let's calculate each component of the formula. Let x = *G.P.A.* and y = *Hours of Radio.*

x	y	xy	x^2	y^2
3.92	4.2	16.464	15.366	17.64
3.78	2.8	0.584	14.288	7.84
2.23	1.5	3.345	4.9729	2.25
2.74	1.7	4.658	7.5076	2.89
3.31	3.5	11.585	10.956	12.25
3.82	4.2	16.044	14.592	17.64
3.02	2.9	8.758	9.1204	8.41
3.17	4.1	12.997	10.049	16.81
2.88	3.6	10.368	8.2944	12.96
2.96	4.1	12.136	8.7616	16.81
3.55	2.7	9.585	12.603	7.29
4.0	2.8	11.2	16	7.84
3.67	3.3	12.111	13.469	10.89
3.31	1.6	5.296	10.956	2.56
3.33	5.8	19.314	11.089	33.64
$\sum_{i=1}^{15} x_i = 49.69$	$\sum_{i=1}^{15} y_i = 48.8$	$\sum_{i=1}^{15} x_i y_i = 164.445$	$\sum_{i=1}^{15} x_i^2 = 168.0255$	$\sum_{i=1}^{15} y_i^2 = 177.72$

Now let's plug the information into the formula:

$$r = \frac{(15)(164.446) - (49.69)(48.8)}{\sqrt{(15)(168.0255) - (49.69)^2}\sqrt{(15)(168.0255) - (48.8)^2}} = \frac{41.818}{\sqrt{51.2864}\sqrt{138.9425}}$$
$$= 0.4954.$$

If we look at the table in the back of the book, we can conclude that, with $n = 15$ and $r = 0.4954$, there is not a significant linear correlation between x and y.

Solution to Practice Problem 10: *Find the linear regression equation of the data set from Practice Problem 7.*

We already have the components of the regression equation from Practice Problem 7:

$\sum_{i=1}^{10} x_i = 1952$	$\sum_{i=1}^{10} y_i = 800$	$\sum_{i=1}^{10} x_i y_i = 159038$	$\sum_{i=1}^{10} x_i^2 = 393538$	$\sum_{i=1}^{10} y_i^2 = 64830$

We can plug these into the formula for the slope. We get

$$a = \frac{10(159{,}038)-(1952)(800)}{10(393{,}538)-(1952)^2} = \frac{28{,}780}{125{,}076} \approx 0.2301.$$

Now we need to find $\overline{x}$ and $\overline{y}$. They are $\overline{x} = \frac{1952}{10} = 195.2$ and $\overline{y} = \frac{800}{10} = 80$. We can plug these into the equation for the y-intercept. We get $b = 80 - (0.2301)(195.2) = 35.0845$.

Therefore, the regression equation is $y = 0.2301x + 35.0845$.

Solution to Practice Problem 11: *Find the linear regression equation of the data set from Practice Problem 8.*

We already have the components of the regression equation from Practice Problem 7:

$\sum_{i=1}^{15} x_i = 49.69$	$\sum_{i=1}^{15} y_i = 40$	$\sum_{i=1}^{15} x_i y_i = 126.482$	$\sum_{i=1}^{15} x_i^2 = 168.0255$	$\sum_{i=1}^{15} y_i^2 = 139.6$

We can plug these into the formula for the slope. We get

$$a = \frac{15(126.842)-(49.69)(40)}{15(168.0255)-(49.69)^2} = \frac{-84.97}{51.2864} \approx -1.6568.$$

Now we need to find $\overline{x}$ and $\overline{y}$. They are $\overline{x} = \frac{49.69}{15} = 3.3127$ and $\overline{y} = \frac{40}{15} = 2.667$. We can plug these into the equation for the y-intercept. We get $b = 2.667 - (-1.6568)(3.3127) = 8.1555$.

Therefore, the regression equation is $y = -1.6568x + 8.1555$.

Solution to Practice Problem 12: *Find the linear regression equation of the data set from Practice Problem 9.*

First of all, it is important to note that we found in Practice Problem 9 that there is not a significant linear correlation between G.P.A. and hours spent listening to the radio. Nonetheless, we will find a regression equation because it is good practice to work with the formula. There may turn out to be a non-linear regression equation that is a better predictor of G.P.A given the number of hours listening to the radio, or that there is simply not much of a relationship.

We already have the components of the regression equation from Practice Problem 7:

$\sum_{i=1}^{15} x_i = 49.69$	$\sum_{i=1}^{15} y_i = 48.8$	$\sum_{i=1}^{15} x_i y_i = 164.445$	$\sum_{i=1}^{15} x_i^2 = 168.0255$	$\sum_{i=1}^{15} y_i^2 = 177.72$

We can plug these into the formula for the slope. We get

$$a = \frac{15(164.445) - (49.69)(48.8)}{15(168.0255) - (49.69)^2} = \frac{41.803}{51.2864} \approx 0.8151.$$

Now we need to find $\overline{x}$ and $\overline{y}$. They are $\overline{x} = \frac{49.69}{15} = 3.3127$ and $\overline{y} = \frac{48.8}{15} = 3.2533$. We can plug these into the equation for the y-intercept. We get $b = 3.2533 - (0.8151)(3.3127) = 0.5531$.

Therefore, the regression equation is $y = 0.8151x + 0.5531$

Appendix

Shaded area = $Pr(Z \leq z)$

TABLE 1
Standard normal curve areas

z	**0.00**	**0.01**	**0.02**	**0.03**	**0.04**	**0.05**	**0.06**	**0.07**	**0.08**	**0.09**
–3.4	0.0003	0.0003	0.0003	0.0003	0.0003	0.0003	0.0003	0.0003	0.0003	0.0002
–3.3	0.0005	0.0005	0.0005	0.0004	0.0004	0.0004	0.0004	0.0004	0.0004	0.0003
–3.2	0.0007	0.0007	0.0006	0.0006	0.0006	0.0006	0.0006	0.0005	0.0005	0.0005
–3.1	0.0010	0.0009	0.0009	0.0009	0.0008	0.0008	0.0008	0.0008	0.0007	0.0007
–3.0	0.0013	0.0013	0.0013	0.0012	0.0012	0.0011	0.0011	0.0011	0.0010	0.0010
–2.9	0.0019	0.0018	0.0018	0.0017	0.0016	0.0016	0.0015	0.0015	0.0014	0.0014
–2.8	0.0026	0.0025	0.0024	0.0023	0.0023	0.0022	0.0021	0.0021	0.0020	0.0019
–2.7	0.0035	0.0034	0.0033	0.0032	0.0031	0.0030	0.0029	0.0028	0.0027	0.0026
–2.6	0.0047	0.0045	0.0044	0.0043	0.0041	0.0040	0.0039	0.0038	0.0037	0.0036
–2.5	0.0062	0.0060	0.0059	0.0057	0.0055	0.0054	0.0052	0.0051	0.0049	0.0048
–2.4	0.0082	0.0080	0.0078	0.0075	0.0073	0.0071	0.0069	0.0068	0.0066	0.0064
–2.3	0.0107	0.0104	0.0102	0.0099	0.0096	0.0094	0.0091	0.0089	0.0087	0.0084
–2.2	0.0139	0.0136	0.0132	0.0129	0.0125	0.0122	0.0119	0.0116	0.0113	0.0110
–2.1	0.0179	0.0174	0.0170	0.0166	0.0162	0.0158	0.0154	0.0150	0.0146	0.0143
–2.0	0.0228	0.0222	0.0217	0.0212	0.0207	0.0202	0.0197	0.0192	0.0188	0.0183
–1.9	0.0287	0.0281	0.0274	0.0268	0.0262	0.0256	0.0250	0.0244	0.0239	0.0233
–1.8	0.0359	0.0351	0.0344	0.0336	0.0329	0.0322	0.0314	0.0307	0.0301	0.0294

(continued)

z	**0.00**	**0.01**	**0.02**	**0.03**	**0.04**	**0.05**	**0.06**	**0.07**	**0.08**	**0.09**
−1.7	0.0446	0.0436	0.0427	0.0418	0.0409	0.0401	0.0392	0.0384	0.0375	0.0367
−1.6	0.0548	0.0537	0.0526	0.0516	0.0505	0.0495	0.0485	0.0475	0.0465	0.0455
−1.5	0.0668	0.0655	0.0643	0.0630	0.0618	0.0606	0.0594	0.0582	0.0571	0.0559
−1.4	0.0808	0.0793	0.0778	0.0764	0.0749	0.0735	0.0721	0.0708	0.0694	0.0681
−1.3	0.0968	0.0951	0.0934	0.0918	0.0901	0.0885	0.0869	0.0853	0.0838	0.0823
−1.2	0.1151	0.1131	0.1112	0.1093	0.1075	0.1056	0.1038	0.1020	0.1003	0.0985
−1.1	0.1357	0.1335	0.1314	0.1292	0.1271	0.1251	0.1230	0.1210	0.1190	0.1170
−1.0	0.1587	0.1562	0.1539	0.1515	0.1492	0.1469	0.1446	0.1423	0.1401	0.1379
−0.9	0.1841	0.1814	0.1788	0.1762	0.1736	0.1711	0.1685	0.1660	0.1635	0.1611
−0.8	0.2119	0.2090	0.2061	0.2033	0.2005	0.1977	0.1949	0.1922	0.1894	0.1867
−0.7	0.2420	0.2389	0.2358	0.2327	0.2296	0.2266	0.2236	0.2206	0.2177	0.2148
−0.6	0.2743	0.2709	0.2676	0.2643	0.2611	0.2578	0.2546	0.2514	0.2483	0.2451
−0.5	0.3085	0.3050	0.3015	0.2981	0.2946	0.2912	0.2877	0.2843	0.2810	0.2776
−0.4	0.3446	0.3409	0.3372	0.3336	0.3300	0.3264	0.3228	0.3192	0.3156	0.3121
−0.3	0.3821	0.3783	0.3745	0.3707	0.3669	0.3632	0.3594	0.3557	0.3520	0.3483
−0.2	0.4207	0.4168	0.4129	0.4090	0.4052	0.4013	0.3974	0.3936	0.3897	0.3859
−0.1	0.4602	0.4562	0.4522	0.4483	0.4443	0.4404	0.4364	0.4325	0.4286	0.4247
−0.0	0.5000	0.4960	0.4920	0.4880	0.4840	0.4801	0.4761	0.4721	0.4681	0.4641

z	**Area**
−3.50	0.00023263
−4.00	0.00003167
−4.50	0.00000340
−5.00	0.00000029

Source: Computed by M. Longnecker using Splus.

z	**0.00**	**0.01**	**0.02**	**0.03**	**0.04**	**0.05**	**0.06**	**0.07**	**0.08**	**0.09**
0.0	0.5000	0.5040	0.5080	0.5120	0.5160	0.5199	0.5239	0.5279	0.5319	0.5359
0.1	0.5398	0.5438	0.5478	0.5517	0.5557	0.5596	0.5636	0.5675	0.5714	0.5753
0.2	0.5793	0.5832	0.5871	0.5910	0.5948	0.5987	0.6026	0.6064	0.6103	0.6141
0.3	0.6179	0.6217	0.6255	0.6293	0.6331	0.6368	0.6406	0.6443	0.6480	0.6517
0.4	0.6554	0.6591	0.6628	0.6664	0.6700	0.6736	0.6772	0.6808	0.6844	0.6879
0.5	0.6915	0.6950	0.6985	0.7019	0.7054	0.7088	0.7123	0.7157	0.7190	0.7224
0.6	0.7257	0.7291	0.7324	0.7357	0.7389	0.7422	0.7454	0.7486	0.7517	0.7549
0.7	0.7580	0.7611	0.7642	0.7673	0.7704	0.7734	0.7764	0.7794	0.7823	0.7852
0.8	0.7881	0.7910	0.7939	0.7967	0.7995	0.8023	0.8051	0.8078	0.8106	0.8133
0.9	0.8159	0.8186	0.8212	0.8238	0.8264	0.8289	0.8315	0.8340	0.8365	0.8389
1.0	0.8413	0.8438	0.8461	0.8485	0.8508	0.8531	0.8554	0.8577	0.8599	0.8621
1.1	0.8643	0.8665	0.8686	0.8708	0.8729	0.8749	0.8770	0.8790	0.8810	0.8830
1.2	0.8849	0.8869	0.8888	0.8907	0.8925	0.8944	0.8962	0.8980	0.8997	0.9015
1.3	0.9032	0.9049	0.9066	0.9082	0.9099	0.9115	0.9131	0.9147	0.9162	0.9177
1.4	0.9192	0.9207	0.9222	0.9236	0.9251	0.9265	0.9279	0.9292	0.9306	0.9319
1.5	0.9332	0.9345	0.9357	0.9370	0.9382	0.9394	0.9406	0.9418	0.9429	0.9441
1.6	0.9452	0.9463	0.9474	0.9484	0.9495	0.9505	0.9515	0.9525	0.9535	0.9545
1.7	0.9554	0.9564	0.9573	0.9582	0.9591	0.9599	0.9608	0.9616	0.9625	0.9633
1.8	0.9641	0.9649	0.9656	0.9664	0.9671	0.9678	0.9686	0.9693	0.9699	0.9706
1.9	0.9713	0.9719	0.9726	0.9732	0.9738	0.9744	0.9750	0.9756	0.9761	0.9767

(continued)

z	0.00	0.01	0.02	0.03	0.04	0.05	0.06	0.07	0.08	0.09
2.0	0.9772	0.9778	0.9783	0.9788	0.9793	0.9798	0.9803	0.9808	0.9812	0.9817
2.1	0.9821	0.9826	0.9830	0.9834	0.9838	0.9842	0.9846	0.9850	0.9854	0.9857
2.2	0.9861	0.9864	0.9868	0.9871	0.9875	0.9878	0.9881	0.9884	0.9887	0.9890
2.3	0.9893	0.9896	0.9898	0.9901	0.9904	0.9906	0.9909	0.9911	0.9913	0.9916
2.4	0.9918	0.9920	0.9922	0.9925	0.9927	0.9929	0.9931	0.9932	0.9934	0.9936
2.5	0.9938	0.9940	0.9941	0.9943	0.9945	0.9946	0.9948	0.9949	0.9951	0.9952
2.6	0.9953	0.9955	0.9956	0.9957	0.9959	0.9960	0.9961	0.9962	0.9963	0.9964
2.7	0.9965	0.9966	0.9967	0.9968	0.9969	0.9970	0.9971	0.9972	0.9973	0.9974
2.8	0.9974	0.9975	0.9976	0.9977	0.9977	0.9978	0.9979	0.9979	0.9980	0.9981
2.9	0.9981	0.9982	0.9982	0.9983	0.9984	0.9984	0.9985	0.9985	0.9986	0.9986
3.0	0.9987	0.9987	0.9987	0.9988	0.9988	0.9989	0.9989	0.9989	0.9990	0.9990
3.1	0.9990	0.9991	0.9991	0.9991	0.9992	0.9992	0.9992	0.9992	0.9993	0.9993
3.2	0.9993	0.9993	0.9994	0.9994	0.9994	0.9994	0.9994	0.9995	0.9995	0.9995
3.3	0.9995	0.9995	0.9995	0.9996	0.9996	0.9996	0.9996	0.9996	0.9996	0.9997
3.4	0.9997	0.9997	0.9997	0.9997	0.9997	0.9997	0.9997	0.9997	0.9997	0.9998

z	Area
3.50	0.99976737
4.00	0.99996833
4.50	0.99999660
5.00	0.99999971

Source: Computed by M. Longnecker using Splus.

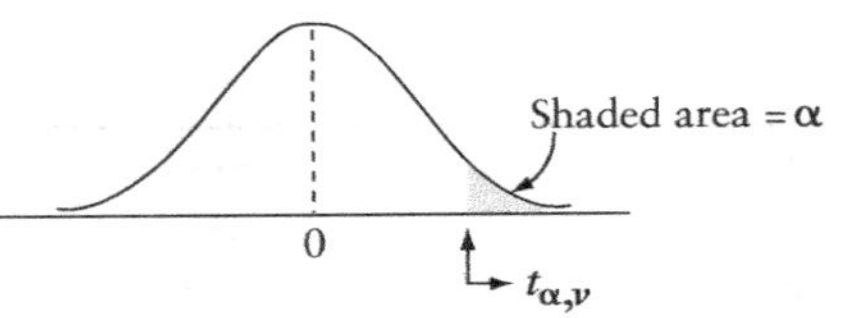

TABLE 2
Percentage points of Student's t distribution

df/α =	.40	.25	.10	.05	.025	.01	.005	.001	.0005
1	0.325	1.000	3.078	6.314	12.706	31.821	63.657	318.309	636.619
2	0.289	0.816	1.886	2.920	4.303	6.965	9.925	22.327	31.599
3	0.277	0.765	1.638	2.353	3.182	4.541	5.841	10.215	12.924
4	0.271	0.741	1.533	2.132	2.776	3.747	4.604	7.173	8.610
5	0.267	0.727	1.476	2.015	2.571	3.365	4.032	5.893	6.869
6	0.265	0.718	1.440	1.943	2.447	3.143	3.707	5.208	5.959
7	0.263	0.711	1.415	1.895	2.365	2.998	3.499	4.785	5.408
8	0.262	0.706	1.397	1.860	2.306	2.896	3.355	4.501	5.041
9	0.261	0.703	1.383	1.833	2.262	2.821	3.250	4.297	4.781
10	0.260	0.700	1.372	1.812	2.228	2.764	3.169	4.144	4.587
11	0.260	0.697	1.363	1.796	2.201	2.718	3.106	4.025	4.437
12	0.259	0.695	1.356	1.782	2.179	2.681	3.055	3.930	4.318
13	0.259	0.694	1.350	1.771	2.160	2.650	3.012	3.852	4.221
14	0.258	0.692	1.345	1.761	2.145	2.624	2.977	3.787	4.140
15	0.258	0.691	1.341	1.753	2.131	2.602	2.947	3.733	4.073

(continued)

df/α =	.40	.25	.10	.05	.025	.01	.005	.001	.0005
16	0.258	0.690	1.337	1.746	2.120	2.583	2.921	3.686	4.015
17	0.257	0.689	1.333	1.740	2.110	2.567	2.898	3.646	3.965
18	0.257	0.688	1.330	1.734	2.101	2.552	2.878	3.610	3.922
19	0.257	0.688	1.328	1.729	2.093	2.539	2.861	3.579	3.883
20	0.257	0.687	1.325	1.725	2.086	2.528	2.845	3.552	3.850
21	0.257	0.686	1.323	1.721	2.080	2.518	2.831	3.527	3.819
22	0.256	0.686	1.321	1.717	2.074	2.508	2.819	3.505	3.792
23	0.256	0.685	1.319	1.714	2.069	2.500	2.807	3.485	3.768
24	0.256	0.685	1.318	1.711	2.064	2.492	2.797	3.467	3.745
25	0.256	0.684	1.316	1.708	2.060	2.485	2.787	3.450	3.725
26	0.256	0.684	1.315	1.706	2.056	2.479	2.779	3.435	3.707
27	0.256	0.684	1.314	1.703	2.052	2.473	2.771	3.421	3.690
28	0.256	0.683	1.313	1.701	2.048	2.467	2.763	3.408	3.674
29	0.256	0.683	1.311	1.699	2.045	2.462	2.756	3.396	3.659
30	0.256	0.683	1.310	1.697	2.042	2.457	2.750	3.385	3.646
35	0.255	0.682	1.306	1.690	2.030	2.438	2.724	3.340	3.591
40	0.255	0.681	1.303	1.684	2.021	2.423	2.704	3.307	3.551
50	0.255	0.679	1.299	1.676	2.009	2.403	2.678	3.261	3.496
60	0.254	0.679	1.296	1.671	2.000	2.390	2.660	3.232	3.460
120	0.254	0.677	1.289	1.658	1.980	2.358	2.617	3.160	3.373
inf.	0.253	0.674	1.282	1.645	1.960	2.326	2.576	3.090	3.291

Source: Computed by M. Longnecker using Splus.

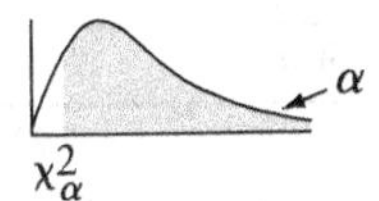

TABLE 3
Percentage points of the chi-square distribution

df α =	.999	.995	.99	.975	.95	.90
1	.000002	.000039	.000157	.000982	.003932	.01579
2	.002001	.01003	.02010	.05064	.1026	.2107
3	.02430	.07172	.1148	.2158	.3518	.5844
4	.09080	.2070	.2971	.4844	.7107	1.064
5	.2102	.4117	.5543	.8312	1.145	1.610
6	.3811	.6757	.8721	1.237	1.635	2.204
7	.5985	.9893	1.239	1.690	2.167	2.833
8	.8571	1.344	1.646	2.180	2.733	3.490
9	1.152	1.735	2.088	2.700	3.325	4.168
10	1.479	2.156	2.558	3.247	3.940	4.865
11	1.834	2.603	3.053	3.816	4.575	5.578
12	2.214	3.074	3.571	4.404	5.226	6.304
13	2.617	3.565	4.107	5.009	5.892	7.042
14	3.041	4.075	4.660	5.629	6.571	7.790
15	3.483	4.601	5.229	6.262	7.261	8.547
16	3.942	5.142	5.812	6.908	7.962	9.312
17	4.416	5.697	6.408	7.564	8.672	10.09
18	4.905	6.265	7.015	8.231	9.390	10.86
19	5.407	6.844	7.633	8.907	10.12	11.65
20	5.921	7.434	8.260	9.591	10.85	12.44
21	6.447	8.034	8.897	10.28	11.59	13.24
22	6.983	8.643	9.542	10.98	12.34	14.04
23	7.529	9.260	10.20	11.69	13.09	14.85
24	8.085	9.886	10.86	12.40	13.85	15.66
25	8.649	10.52	11.52	13.12	14.61	16.47
26	9.222	11.16	12.20	13.84	15.38	17.29
27	9.803	11.81	12.88	14.57	16.15	18.11
28	10.39	12.46	13.56	15.31	16.93	18.94
29	10.99	13.12	14.26	16.06	17.71	19.77
30	11.59	13.79	14.95	16.79	18.49	20.60

(continued)

df α =	.999	.995	.99	.975	.95	.90
40	17.92	20.71	22.16	24.43	26.51	29.05
50	24.67	27.99	29.71	32.36	34.76	37.69
60	31.74	35.53	37.48	40.48	43.19	46.46
70	39.04	43.28	45.44	48.76	51.74	55.33
80	46.52	51.17	53.54	57.15	60.39	64.28
90	54.16	59.20	61.75	65.65	69.13	73.29
100	61.92	67.33	70.06	74.22	77.93	82.36
120	77.76	83.85	86.92	91.57	95.70	100.62
240	177.95	187.32	191.99	198.98	205.14	212.39

α = .10	.05	.025	.01	.005	.001	df
2.706	3.841	5.024	6.635	7.879	10.83	1
4.605	5.991	7.378	9.210	10.60	13.82	2
6.251	7.815	9.348	11.34	12.84	16.27	3
7.779	9.488	11.14	13.28	14.86	18.47	4
9.236	11.07	12.83	15.09	16.75	20.52	5
10.64	12.59	14.45	16.81	18.55	22.46	6
12.02	14.07	16.01	18.48	20.28	24.32	7
13.36	15.51	17.53	20.09	21.95	26.12	8
14.68	16.92	19.02	21.67	23.59	27.88	9
15.99	18.31	20.48	23.21	25.19	29.59	10
17.28	19.68	21.92	24.72	26.76	31.27	11
18.55	21.03	23.34	26.22	28.30	32.91	12
19.81	22.36	24.74	27.69	29.82	34.53	13
21.06	23.68	26.12	29.14	31.32	36.12	14
22.31	25.00	27.49	30.58	32.80	37.70	15
23.54	26.30	28.85	32.00	34.27	39.25	16
24.77	27.59	30.19	33.41	35.72	40.79	17
25.99	28.87	31.53	34.81	37.16	42.31	18
27.20	30.14	32.85	36.19	38.58	43.82	19
28.41	31.41	34.17	37.57	40.00	45.31	20
29.62	32.67	35.48	38.93	41.40	46.80	21
30.81	33.92	36.78	40.29	42.80	48.27	22
32.01	35.17	38.08	41.64	44.18	49.73	23
33.20	36.42	39.36	42.98	45.56	51.18	24
34.38	37.65	40.65	44.31	46.93	52.62	25

(continued)

$\alpha = .10$	.05	.025	.01	.005	.001	df
35.56	38.89	41.92	45.64	48.29	54.05	26
36.74	40.11	43.19	46.96	49.65	55.48	27
37.92	41.34	44.46	48.28	50.99	56.89	28
39.09	42.56	45.72	49.59	52.34	58.30	29
40.26	43.77	46.98	50.89	53.67	59.70	30
51.81	55.76	59.34	63.69	66.77	73.40	40
63.17	67.50	71.42	76.15	79.49	86.66	50
74.40	79.08	83.30	88.38	91.95	99.61	60
85.53	90.53	95.02	100.43	104.21	112.32	70
96.58	101.88	106.63	112.33	116.32	124.84	80
107.57	113.15	118.14	124.12	128.30	137.21	90
118.50	124.34	129.56	135.81	140.17	149.45	100
140.23	146.57	152.21	158.95	163.65	173.62	120
268.47	277.14	284.80	293.89	300.18	313.44	240

Source: Computed by P. J. Hildebrand.

TABLE 4

Critical Values of the Pearson Correlation Coefficient r

n	$\alpha = .05$	$\alpha = .01$
4	.950	.999
5	.878	.959
6	.811	.917
7	.754	.875
8	.707	.834
9	.666	.798
10	.632	.765
11	.602	.735
12	.576	.708
13	.553	.684
14	.532	.661
15	.514	.641
16	.497	.623
17	.482	.606
18	.468	.590
19	.456	.575
20	.444	.561
25	.396	.505
30	.361	.463
35	.335	.430
40	.312	.402
45	.294	.378
50	.279	.361
60	.254	.330
70	.236	.305
80	.220	.286
90	.207	.269
100	.196	.256

NOTE: To test H_0: $\rho = 0$ against H_1: $\rho \neq 0$, reject H_0 if the absolute value of r is greater than the critical value in the table.